MW01485712

On Top of The World

A Life's Education Through

Travel & Adventure

Table of Content

Acknowledgments

I would like to take this opportunity to acknowledge the people who helped make this book a reality for me. As I have stated many times before, a person's success is never the result of something they do all on their own, but the result of many people who help that individual reach their goal.

- As with all my books, I will always dedicate my work to my daughters Alissa, Katie & Megan first. They are my inspiration, and my reason for wanting to continue to reach new heights. It is my pleasure to share many stories within this book that includes them with you.
- I would also like to thank all the people whom I met along the way and were part of my adventures. Each of these individuals helped make every story more memorable for me and played a role in my growth and development as a person.
- Finally, I wish to thank all of the people who have supported my writing and work the last several years. When there were times I felt like I had nothing more to say, they reminded me that there is always something more to say.

Take Off

"Ladies and gentlemen, the forward door has now been closed, and we are ready for take-off. At this time all electronic devices must be turned off and stowed under the seat in front of you, and remain there until the Captain has advised that they may be used again"

I have memorized some variation of that speech having heard it literally thousands of times in my travel for work and pleasure. Hearing it that many times, I can almost repeat it verbatim depending upon which airline I am flying, but the sound of it never gets old. For that speech means that I am embarking on another adventure of traveling to a place somewhere besides my home. Let me say right off the top, I have been blessed to work in an industry that allowed me to travel for work, and in many cases, travel for pleasure and adventure as well. Now don't go kicking yourself because you think you chose the wrong profession for all these years, as glamorous as traveling around the globe can be, it comes with a price to be paid, like any other job.

When I first entered the travel industry over 30 years ago, it was a well-known fact that salaries in the industry were below average, but the travel benefits available helped make it a desirable pro*fession to enter. Somehow* at 21 years old, I had the wisdom to think I would enjoy the experiences at that time, more than making a good salary. A wisdom that I would eventually forget during my formative 30's and 40's, only to return in my 50's. Indeed, I was able to take advantage of those significant discounts in the early days, but much like anything else, less has become more, and in today's travel industry, the advantages that made the experiences possible, have all but gone completely away. There have been instances where being a savvy traveler using loyalty programs, has far exceeded any travel perks I may have been entitled to.

Whether it be travel perks for personal travel, or the requirements for business related travel, I have always tried to maintain one philosophy. Wherever I was traveling to, I would try to make the most of that trip, and experience something new. Regardless of the length of time, or size of the city, there was always something new to discover. Not everyone has the same philosophy as I do, in fact most of the colleagues I have traveled with over the years, did not have the same zest for adventure that was bursting inside of me. I didn't matter if it was **Harry Potter** World in Orlando, the **Mary Tyler Moore** statue in Downtown Minneapolis, or the last of the White Buffaloes in the **Badlands of North Dakota**, if there was a way for me to check it out, I made sure to do it.

It is that spirit that I wish to share with you, and the impact it has had on my life. No matter where it was that I traveled to, I learned to take away some new knowledge of where I had been. Adventure travel does not necessarily mean some exotic far away location that you only dream about, or some location that everyone of your friends has visited. It can also encompass a small village in your home state, or whatever gives you the traveler pleasure. I talk to people all the time who feel that the only way to experience a new place is to travel the way a local person does. That is nonsense, as many trips I have taken have been that of a typical tourist, and some have been as a native. The true measure of your travel is what you want to get out of the experience, and not necessarily where you go. I have traveled to exotic destinations, and been very disappointed, while also traveling to places I had serious reservations about, only to come away with a sense of accomplishment that no brochure could ever provide. It is because of that mindset that I have truly been on top of the world.

During my career, I have had the opportunity to visit 5 of the 7 continents on this planet, and still actively working for 6 and 7. If you are one who uses the "Bucket List" concept, I guess reaching all 7 continents is a bucket list item for me. I also have a goal of seeing the 7 natural wonders of the world, a goal that I believe is very achievable, as I currently have crossed off 4 of the 7. While I have these lofty goals, it is just as plausible that I won't be able to see everything that I outline on my "to do" list. That is not a failure, because I will always still have that dream, and if we don't dream big, then it really isn't a dream is it?

What my goal here in this book is to share each of the experiences I had, and also share how the experience energized and change my life in some way. Whether it was a change of perspective, an appreciation of what I have, a love of a new foods, or an awakening of empathy for others that was lying dormant in my soul. Each of the trips I write about, has made some lasting impact on my life. My hope is that by sharing some of this with you, a spark will start within your life to experience new adventures and create memorable moments in your life. The stories will not be in a chronological timeline but be told based on geographic location. Therefore, some of the stories you read first, may be more recent than others, but each will be put into proper time perspective. There will be stories where I share some financial information in terms of discounts, freebies, and relative costs I may have incurred. I do so not as a means of bragging, but rather to demonstrate that the benefits I received in travel, made most of these trips possible.

I have also included a section on what I call local adventures, because leading an adventurous life doesn't always mean you have to travel the globe to do so. In fact, some of the most adventurous things I have done, have been right in my own back yard. Every one of us has adventures in our life that we don't even think of as adventure, but what we need to remember is that "Life" is an adventure in itself. So, much as I did in my last book "Where U At", I want you to think about adventure in a different way, and I am convinced, that you will identify things from your past, that were adventurous, and that it will motivate you to continue this newly discovered passion.

To be clear, I am not going to document every single trip, or adventure I have had in my life, but I am going to focus on the ones that I feel were fundamental in my approach to how I live, and what is important to me. Along with that, I will share why I believe it was important to me, and perhaps what changed within me, that made it worthy of sharing with all of you. There is no doubt in my mind that what I have seen and done, has had a profound impact in my life. It has made me think about things I probably never would have without the experiences. Experiencing new things has given me the energy to continue seeking new and different outlets, while increasing my knowledge of the world. So, whether you travel far or near, or whether you seek adventures, I urge you not to wait, and get out and seek your passion.

Thank you for taking this trip with me and allowing me to show you the world in my eyes.

Let me take you on a trip
Round the world and back
And you don't have to move, you just sit there
Let me show you the world in my eyes
(Depeche Mode)

1-Australia

Probably the one place that people have often said to me in conversations that they have always wanted to visit would be Australia. In my conversations about travel, I would mention that I had been there back in 1992, and many people were often surprised by my rather "**no big deal**" take on the trip. I will get to that more in this chapter, but first the outline of how I ended up on the other side of the world. It's no secret that Australia is somewhat a mythical place for many people in North America. I think the sheer distance between there and America has a little to do with it, but I also think some of the more famous native animals to Australia, provide an intrigue that draws people to make the journey. For me, it was not a place that I had on my early radar to visit, but as you read on, life can take some mysterious turns.

One great advantage of traveling globally is that you often meet people from other parts of the world. From those meetings, friendships are often born, and so it was that a friendship that was made 2 years earlier in Europe, allowed me to make another journey in my young life. Having a friend in another country makes for a much more affordable trip, especially when they provide accommodations for you to stay at. This was a big factor to entice me to make a trip to visit my friend **Lindsay** in the land that is not only a country but is its own continent as well. You will read more about Lindsay in detail in a coming chapter, particularly when we get to Europe, but for now, it was time for him to show me his country.

A second advantage is to have a job that affords you the opportunity to receive great travel benefits to offset some of the costs of a significant trip such as this. During this time, I was working with American Express, and of course because they have such huge buying power, many vendors would often present incentives to us, in order to secure business. So, when **United Airlines** came a calling in early 1992 to offer us a sales incentive contest, the dye on my next trip was cast. The travel industry is not unlike other industries where an individual will support another product that often supports them. I had a significant amount of persuasion in dealing with the general public with regard to selling airline vendors. In other words, whoever took care of me, would receive the same reward in kind. For a while, **American Airlines** had probably been my preferred vendor to sell, as I had some nice perks early in my career from them, and I wanted to support them. But when United decided to offer an unlimited sales incentive, it was very easy to switch my selling techniques, and paint them in a more favorable light. So, because our office did really well with the contest, I ended up with 2 free tickets that would jet me across the Pacific to another new location. Australia here we come!

My wife Mary and I decided to plan the trip in August to coincide with her Summer break from teaching, but a couple of items had us rethinking the strategy. First off, Australia is in the Southern Hemisphere, so their seasons are the reverse of ours. August is actually their Winter, and some parts of the country would be experiencing Winter type weather. A second and more important item, was that in July of 1992, we were greeted with the news that our first daughter Alissa would be joining our family in February of 1993. Taking a journey of this magnitude with a first-time mom would not be easy, and we had some concerns. We knew the tickets had an expiration date, and if we had waited until after the baby was born, the chances of us taking the trip, would have been miniscule. So, we stayed on schedule and took our trip in late August of 1992.

By this time in my life, I had flown on some pretty long flights, but this would be my longest nonstop flight, and it was by far the longest for Mary as well. Fourteen and half hours from Los Angeles to Sydney, and in coach class, would not be very comfortable for a 10-week pregnant lady. We managed to make the best of it, and finally arrived in Sydney, beyond exhausted 2 calendar days later. Lindsay met us at the airport, and immediately took us to his house where his mom Wendy was awaiting our arrival. An amazing lady Wendy was, she is no longer with us, as she had a hot water bottle ready for Mary's arrival knowing her condition. It was that preparedness that allowed us to get some much-needed rest before starting on our sightseeing adventures.

For our first day of sightseeing, Lindsay took us to an area known as the Blue Mountains. Again, because we were in the Winter season, the **Blue Mountains** were actually quite cold, and required that we have coats. The scenery was very nice, and in many cases, looked a lot like any other mountain community. The big scenic draw was the **Three Sisters**, which is 3 large rock pillars next to one another along the mountain side. This day was a little less hectic given that we were still acclimating from the travel, and Mary's pregnancy. It was a good first day, but now it was time to get to the things that we really came to see.

Featherdale Wildlife Park had what we were looking for with Koala Bears, Kookaburra's and of course, Kangaroos. We actually got to hold a Koala Bear for picture purposes with supervision, as Koala's while seeming very gentle, can have a mean streak in them if provoked. What I found so impressive was the intelligence of the Kangaroos, and their willingness to come up to humans to be fed and petted. Much like a household dog, kangaroos will come when you call them, and can be very affectionate. Also, of keen interest was the way they use their very strong tails as a direction guide such as turning around. Overall it was great to be able to experience some of the native animals of Australia, but the big daddy still was yet to come for me.

Mary & I feeding a baby kangaroo in Sydney, August 1992

The city of Sydney is much like many big cities in America such as New York or Chicago. There is mass transportation, a lot of traffic, and many buildings. We made our way down to see the famous **Sydney Opera House**, which is pretty much as you have seen it in pictures. Sitting along the harbor, it can be seen from many different viewpoints, and is within view of the spectacular Harbor Bridge. The Harbor Bridge has very much of a **Brooklyn Bridge** feel to it with a very similar look and design. Some other points of interest in the city were Darling Harbor, and Circular Quay, (pronounced key). These are very popular places for eating, shopping, and just getting out to enjoy the day. We finally ended up in the area known as **The Rocks**. This is the old part of the city, as it was the original start of the city of Sydney. The area also holds the infamous distinction of being the criminal settlement, as Australia was originally settled as a penal colony of the British. Whether it was mountain scenery, exotic wildlife, or the modern city with the infamous past, our visit to New South Wales proved to be a great start to this vacation. Now it was time to see just how much beach this island continent had, and the opportunity to see one of the 7 natural wonders of the world.

The Sydney Harbor Bridge, August 1992

QANTAS

Most people recognize the name of Australia's flag carrier; however, many do not know what it actually means. The name **QANTAS** is an acronym for *Queensland and Northern Territory Aerial Services*. It was Queensland that was next on our list for travel on our trip of Australia, more specifically, the city of **Cairns** (pronounced cans). Up until this time, our trip had been fairly traditional in that we had seen some native animals, great mountain scenery, and the metropolis of a big city and its architecture. Driving on the left side of the road is always challenging, but I had actually experienced that before on my first trip to Great Britain a few years earlier. Nothing was really unexpected per se, but that would soon change. Queensland is a state to the North of New South Wales, and because we are in the Southern Hemisphere, heading North actually means you are heading closer to the equator, and therefore, the weather is warmer. The city of Cairns is a kind of jumping off point to the **Great Barrier Reef,** where one can experience some of the best snorkeling and diving in the world.

As we boarded our flight from Sydney, the process of boarding and take off was standard, but about 30 minutes into the flight, I knew I was no longer in America. I had been on many flights where prior to take off, the cockpit crew would often allow a few people, mainly children to come up and look inside the cockpit. The amount of controls, buttons, and lights is amazing, so I was always fascinated to come up and look as well. On this flight however, after a cruising altitude was reached, an announcement was made that any kids were welcome to come up and see the cockpit. As my hand was part way up, I waited to see if any young ones volunteered. After about 20 seconds, my hand went fully up, and I dragged Mary and Lindsay along for the fun. Actually, getting to experience the cockpit from 30,000 feet was something to behold, and something that the FAA in America never allowed. The flight crew took the time to explain some of the instrumentation, and actually told us where in the flight plan we were. The plane essentially flies itself, and that became evident as we flew over a city called **Townsville**, and the plane automatically started into a slow descent for our eventual arrival in Cairns. It was the only time I have ever been in the cockpit of a flying airliner, and in a post 9/11 world, most likely the only time I ever will.

Arriving in Cairns was much different than our arrival in Sydney, as the cool temperatures gave way to the hot, humid tropical weather. This is where we would experience the tropical island climate that so many people long for all year for a vacation. Our first day there, had us checking out the **Kuranda Rain Forest**. Which essentially means it is like any other rain forest, but does have a cool railroad that runs right through it. We were able to take this train ride through the forest all the way down the mountain. The Cairns area is also overrun with sugar cane fields, as it is a significant supply of sugar for the entire country.

One interesting aspect about the architecture of the area is that given the potential for tropical storms, houses are built on stilts to help prevent flooding. This was present even for houses that were away from the water line and considerably inland.

The afternoon saw us head to one of my top highlights of the whole trip as we finally arrived at **Hadley's Crocodile Creek.** A crocodile farm strictly dedicated to preserving one of Australia's main inhabitants. Like many people, I have both an aversion to reptiles, and an utter fascination with these creatures. This is the top of the food chain, and its lineage can be traced all the way back to the Jurassic period. Big, small, fresh water, and salt water crocs can be found at Hadley's. One of the shows we watched had a **Steve Irwin** type guy actually feeding the crocodile as part of the show with dead chickens, fur and all. Standing beside a 15-foot croc, talking to the crowd, and showing no signs of fear. It became clear to me that the little kid in me came to the surface with both fear, and excitement, but now it was time for a bit of sun and fun.

The next day had us flying across the waves in a hovercraft as we took a snorkeling day trip with **Quicksilver tours.** The tour would take us to a section of the Barrier Reef, and then park for several hours to partake in snorkeling. I had purchased an inexpensive underwater disposable camera so that I could take photos of the reef, and marine life we would see. The water was amazingly clear, and the colors of some of the fish that I saw, was simply stunning. I learned much about the reef, with the most important aspect that the reef itself a living thing and supports an essential ecosystem for all the marine life there. It is said that the Great Barrier Reef is one of the few things that can be identified from space, which does not surprise me given the massive size overall. Lunch was provided on the tour which consisted of sandwiches, salads, and some seafood. It was here that I mistakenly tried one of Australia's famous sandwiches, vegemite!! That's right, the sandwich made so famous in the **Men at Work** song "**Down Under**" was right there for the taking, and I foolishly took it. Lindsay had told me that vegemite tasted like peanut butter, but he also had an axe to grind, as I convinced him to try pumpkin pie on Thanksgiving a year earlier. One bite into this thing had me reaching for the napkin with the pretend cough back into the napkin. It was not peanut butter, and it was not good, but it did provide a laugh for Lindsay and my wife.

Our final day in Cairns was reserved for a day at the beach, just to relax and enjoy the Pacific Ocean from the other side of the world. The nice thing about the time of year that we went, was that the **Box Jelly Fish**, which migrates in their Summer, was not in season, so this allowed us to swim in the ocean. It is hard to imagine that in many parts of the Summer, you cannot swim in the water because of the potential for jelly fish stings. As we departed Cairns for our return back to Sydney, I looked back at what we had seen the last couple of days, and savored it, knowing I may not have the opportunity to return.

I'm sure a lot of people have experienced a flight that has been canceled for some reason, such as weather or mechanical. Those are mild inconveniences, and the airline usually has you on your way on the next flight or so. It's not quite the same on an international flight across the world, as we found out when United Airlines called us the morning of our departure to advise us that our flight home had been canceled. In this instance, there wasn't another plane that just happens to be there leaving an hour later. We ended up having to wait 15 hours before another plane would arrive for us to head back to America. This unexpected extra time gave us the opportunity for some additional city sightseeing and gave me the chance to take some excellent photos of the Harbor Bridge, and Opera House. An afternoon lunch with Lindsay and Mary along the Harbor, allowed for one last chance to have a **Thuey's Beer**.

Don't remember which one of the 8 varieties that I had, but a last chance to take in something uniquely Australian. For every McDonalds, there was a Thuey's, for every restroom, there was a toilet, and for every hamburger, there was a shrimp on the Barbie. As we departed for what would be 12:30 minute flight home, yes the tailwind makes up 2 hours, I remember looking out at the night Sydney sky, and thinking would I ever see my friends again, or was this the finishing piece where we each visited one another's home, and that's it? I eventually learned that true friendships can outlast any distance, even one that is 7500 miles away.

What's left?

Given the vast size of Australia, and the relatively short vacation time that we had, there were a couple of things that we could not fit into the trip, and I had slated for the next time.

- The Outback: particularly Alice Springs where the Outback meets civilization
- Ayers Rock: Given my propensity to climb things, this really should have been a must, but getting there is an adventure itself. Being able to climb the Rock is a Bucket List item for me.
- Phillip Island: Experiencing the Penguin Parade back to the sea would be an amazing thing to witness. Until then, I will have to settle for the Pittsburgh Penguin Parade.

2 – North America
(INCLUDING CARIBBEAN)

It probably isn't a big surprise that this chapter just might be the longest in the book, and that I have listed the greatest number of monumental trips in this chapter. It would seem appropriate given this is where I live, and the relative proximity to most of these trips is within a few hours travel time. There are two important elements that I want to point out about my North American travels. The first is that a trip does not have to be half way around the world in order to hold a meaningful significance. It is just as likely that a person can have an epiphany on a week end getaway 50 miles from home. The second thing is that many of the trips I highlight in this chapter, coincide with where I was in life in relation to growing kids, and family. Whether these trips were big or small, each of them held some turning point for me, and played a role in my life experiences.

Ft. Lauderdale, FL
MARCH 1985

The very beginning of my passion for travel can be traced back to the Spring of 1985. It certainly was not the first time I ever traveled, because I had taken many vacations with my parents before this, but this was the very first vacation that I took on my own. Not just traveling by myself, but every aspect of planning and executing the trip was handled by me. I was able to pay for my own airline ticket, purchase some new clothes for the trip with the money I earned from my job, and ultimately plan the dates without any assistance from family. So, in many ways, it was a liberating experience into adulthood, even though I was only 20 years old at the time.

The reason for this trip was a chance for me to experience **Spring Break**, in what was known as the Spring Break Capitol at that time. It just so happened that my buddy Greg moved down to Ft. Lauderdale a few months earlier to work with his brothers. As I will mention throughout the book, it really helps when you have friends in places that offer you a place to stay for free. I knew what Spring Break was, and what it symbolized, but had never experienced anything like it before. I was still a very shy individual at this time, and really needed to break out of my life shell. This seemed like the perfect opportunity to come into my own, so in early March, I took my very first adult vacation.

The arrival my first night was pretty crazy, and much of what I had seen in the movies was playing out verbatim. College kids and parties everywhere, and yes, a lot of alcohol and sheer craziness as well. This was so different from what my everyday life was like, and a little scary, but I was determined to join the world, and leave behind the shy kid, who couldn't even speak to women. The weather for this week was amazing for the beach in the daytime, and clubs etc. at night. One very important thing that helped contribute to my fun was that the drinking age in Florida at this time was 19. This was before the country followed the universal age 21 standard, so I was able to enter clubs, and have alcoholic drinks legally. No more worrying about fake ID's even though I wouldn't be 21 for about 5 more months. Each night there was something new to do, and a chance to let loose, but still maintain some element of decorum. It was that letting loose where I experienced my first female mud wrestling contests. Prize money, and a whole lot of fun for the women who would challenge one another in a big pen of mud.

The other draw of Spring Break is with the plethora of bands that play the clubs in the area during this time. Not just local cover bands, but many popular bands of the day like **Autograph and Survivor** were there as well. I remember meeting some guys from a Pennsylvania band called "Magnum" that I liked a lot. Having the Pennsylvania connection allowed us to strike up a friendship, and I watched their show a few different nights. It was there at the **Playpen South** that it happened, on the other side of the room, I saw her. This cute girl hanging close to the band. Her name was Lisa, and we hung out for the next several days. A real-life vacation romance took over, and at the end of the week, it had to end. Though I was convinced that this girl from Jersey, and a love-struck guy from Pittsburgh could make a go of it.

Much sooner than I would have liked, my wild Spring Break week came to an end, and it was time to board the flight home. It was 85 degrees when I left and snowing when I arrived back in Pittsburgh. I had a pretty good tan going, and got off the plane in shorts, but the memories were still back in Florida. Though it was hard to leave, I came home knowing that I now experienced my own vacation, my own fun, and my own responsibilities. I knew that I would take other trips, but still didn't have any idea how prominent traveling would become in my life.

Killington, VT
JANUARY *1986*

After I returned from Spring Break, I gave some serious thought to this whole travel thing and decided to find out about getting more involved with it. I made friends with the people at the local travel agency in the mall and started to learn the computer system. Then I took a course on **Travel Agentry** through extended education, and by the end of 1985, convinced the Travel Agency owner to give me a try. This new job coincided with something that I had wanted to do for a long time.

I had been skiing for several years now, but it had only been at local Pennsylvania and New York resorts. I had wanted to give one of the famous New England resorts a try, the kind you would hear about on the news. It was hard to find people that were as passionate about skiing as I was, but I managed to convince a couple of buddies to take a trip to Killington, VT and try some real skiing for a change. This was a full ski vacation in that we would fly to Vermont, rent a car, and drive to the resort. There was a total of 5 guys who would make this trip, and on the first night arrival in Burlington, VT, we shared one hotel room for the night. For me, this trip was about traveling with a group of friends, and again proving to myself that I was now a man and could make adult decisions. I was never one to be affected by peer pressure; however, I was also trying to shed an image of being afraid and boring, so it was on this first evening that I would have my first and only experience with cocaine. I had always been curious as to the attraction, and the popularity of cocaine, so this was my chance to try it out. 2 lines later and feeling no different, I didn't understand the attraction, let alone the expense that was involved with this attraction. Having given it a try, I decided that is was no big deal, but did become hooked on Coca Cola later in life. This allowed me to honestly tell people that I have a bad Coke problem.

The next morning was Super bowl Sunday, and we drove from Burlington down to Killington in the morning, so we would be able to watch the game. The game itself wasn't much as the 85 Bears blew out New England, but it was an early night as tomorrow was the first day of Vermont skiing. Our first day the conditions were really good, as this was Vermont powder we were dealing with, not the usual manmade snow of **Seven Springs.** The terrain was made for Intermediate to advanced skiers, but they had a 2 ½ mile leisurely run that went from the top of the mountain, all the way down. It was a great day of skiing, but it was time for us to check out the local night life. We came across a place in town called the **Night Spot,** which was a local dance club. The place was inhabited with college girls from **Princeton University** who were on semester break. The club was really hopping, and just like that, there she was, another vacation fling. This young lady was another Lisa, who was from Scotland. Her accent alone was enough for me to swim across the Atlantic Ocean to see her again. This turned out to be only a one evening encounter, and was not as detailed as my Spring Fling, but of course I left it with me telling her how I would get to Scotland one day. I had no idea how prophetic that statement would become 10 months later.

The next day was January 28, 1986, and if you are wondering why I emphasized that date, it is because on that day, the **Space Shuttle Challenger** exploded. If I was ever asked "where was I when the Challenger exploded", the answer is Killington, VT. To make matters worse, Killington was experiencing a white out that day, and skiing was closed because the visibility was nothing, and the wind was severe. All we were able to do was hang out in the hotel and watch the news over and over again showing the video of the explosion. I learned more about Solid Rocket Boosters than I ever cared to. A very sad day, but one that is etched in my mind forever. Aside from the disaster, the trip was a success from all other aspects, and one that provided me with some firsts, even if some of those firsts were not my finest moments. It was all part of growing into a adult and growing on my own.

Jamaica
JULY 1986

By July of 1986, I had about 6 months under my belt as a full-fledged Travel Agent. I really enjoyed the work, and had a great team that I worked with, so now it was time to experience some unique travel. It wasn't often that free airline passes go to waste, but it just so happened that our office had a pass on Eastern Airlines (remember them?) that was about to expire if not used. Me being the new guy had the first opportunity to make use of the pass. The pass was good for anywhere in the USA or Caribbean, so there was no way I would waste this pass on some USA destination, but where to go? A friend of mine told me about a place in Jamaica where everything was included in one price. The all-inclusive vacation is a very common concept nowadays, but back in 1986 was still relatively new. The name of the club was **Hedonism II,** and as I learned more about it, the name certainly fit the property. I called the club, and discovered they offered a great agent rate, and booked myself for 4 nights. I had heard the place was pretty risqué, but thought it was just some exaggerated tales from people based on what they read, but soon I would find out for myself.

As with most of my trips at this time, there continued to be many first's. This would be the first time I actually traveled outside the country, and also the first time I had taken a flight over a body of water. Another first was experiencing the concept of Space Available airline travel. This is basically a stand-by situation where if there is an unused seat, you can have it, but if the plane is full, you can be stranded, and not get on the flight. This is one of the drawbacks of free airline travel and is always a possibility when you travel on industry discounts. Fortunately for this trip, it was an easy connection through Miami, continuing on to Montego Bay, Jamaica.

The arrival at the airport in Jamaica went smoothly, except for a gentleman who came up to me and tried to put what appeared to be some sort of drugs in the palm of my hand. You often hear horror stories of people trying to slip illegal items into your person, and that was my immediate thought. After a long shuttle ride, we arrived at the club in the city of Negril, on the West side of the island. When I checked in, the front desk handed me a couple of bed sheets, and my immediate thought was "I have to make my own bed, how lame is that". It turned out the sheets they gave us were for the Toga party that was taking place that night in the main hall. My first night in and already a toga party was underway, along with unlimited drinks and reggae music. I started to meet people through the course of the night, and they were from all over the world, yet gathered together in white sheets having a good time. Besides the mischief, the club also offered every type of watersports you could think of, from scuba & snorkeling, to wind surfing & nude sun bathing. My room just happened to be along a stretch of the beach that was reserved for all-natural bathing, so consequently, waking up to moon, much more than sun.

There was also an opportunity to take some excursions outside of the club as well. One trip was to a small bar near the tip of the island called **Rick's Café.** Nothing really too elaborate about the place, except for a pedestal for cliff diving with a 30-foot drop into the Caribbean. Soon the people started jumping off the rock, and this became my first opportunity to challenge my fear of heights. I remember literally shaking as I approached the small platform, and for just a split second I thought I might chicken out. Having summoned up all my courage, I leaped outward as far as I could, and did the 30-foot plunge down into the sea. It was awesome to try, but I only did it the one time. However, that jump started a process that continues today with me challenging that ever constant fear of heights.

On the way back from Rick's, there was also a planned stop at a place called **Jennie's Famous Cakes.** This was nothing more than a roadside shack made famous for pot brownies and mushroom tea. Still being wet behind the ears with the drug culture, I knew what pot brownies were, but **shroom tea** was something I was not familiar with. I had a few tastes of it, and it wasn't all that great, but I understand if you drink enough of it, it will produce a hallucinogenic state, and taste becomes less of a priority.

I had one more day to enjoy this island lifestyle, and that consisted of snorkeling right from the beach. All was fine right up until I came across a sea snake swimming by, and then it was back to the bar for a banana daiquiri. The flight home was pretty much normal with the massive thunderstorm in Miami, and a 2-hour delay, but after the fun and experience I had the last 4 days, what's a little delay. I remember getting on the plane in Miami thinking how I could get used to this traveling stuff, and how I had a great vacation for about $250.00 dollars. Did I really fall into an amazing career, or would this just be a season in my life?

St. Maarten
APRIL 1987

Now that I had a year and half of travel agent experience under my belt, along with experiencing travel and the perks it came with, I started to settle into my own. I had been fortunate with many different trips, and I thought it was time to give something back. Due to an Eastern Airlines ticket incentive, I had a couple of free Caribbean tickets to use, and these were Positive Space, which means confirmed seats regardless of the plane capacity. I wanted to do something special for my mother, who was very patient with me while trying to find myself, and a career, so I decided to let her experience the travel I had experienced and took her to an island in the Dutch West Indies, St. Maarten. She had always wanted to go to Hawaii, but I thought she might enjoy something similar to that tropical climate. We showed up at the airport for our flight down to San Juan, and right away the perks kicked in as Eastern upgraded us to first class. Eastern Airlines was always the best at upgrading travel agents. So right away, I was a big shot with my mom having her fly first class, and a Pina Colada in her hand before the plane even took off.

That pretty much was the highlight of the trip because from there on out, I was forever reminded that I was traveling with my mother, and it was no longer about what I wanted. We traveled to a Caribbean island, stayed right on the beach, and she was insisting we rent a car, and go jewelry shopping in the Downtown area. One store after the other, "look at this ring", "isn't that ring pretty", until I was pretty much at the end of my rope. To make matters worse, the dates we chose coincided with national Dutch holidays, and many stores were closed. Of course, I was reminded every time we went to the beach that the stores were closed, and there was nothing to do. On one of the days, I took her for a drive around the island, and over to the French side of the island, just so I would not have to listen to her.

This was probably the longest 4 days of my life, only to be capped off on the flight home as she complained to the passenger next to her about what a lousy time she had because of the holidays etc. When the passenger told her, she should complain to her travel agent about the trip, her response was, "I can't, he's my son". At this point my takeaway lesson was no more traveling with mother, and that traveling alone, certainly has its benefits. A concept that to this day still holds true in many instances. This was probably the first time I had what I would consider to be a negative traveling experience.

Hawaii
July 1987

It was never really a question of "if" I would get to Hawaii, but more like "when" I would have the opportunity. Up to this point, the situation never really presented itself for me to head that far West. But several things were in the works that would change that sooner than I had projected. I started to get restless in my career and would soon have to make some decisions. I loved my job but making $5.25 an hour was not allowing me to strike out on my own, and I began to realize that if I wanted to have my own place, I would need to make more money. I had visited California several times, and always liked the atmosphere, and of course the weather. So, I decided to take some vacation time, and look for a job in California, and because I was there, I would add a few days in Hawaii for my first trip to our Country's 50th state.

I was very fortunate that after a couple of days in California, and going on several interviews, I was offered a job. It wasn't a spectacular offer, but it would pay enough for me to get by, and start a new life, in a new home. Once that part of the trip was successful, it was time to continue on to Hawaii, and a day earlier than expected. Again, taking advantage of some industry promotions, I had free accommodations at the **Hawaiian Regent Hotel,** and some other discounts on activities once I got there. The hotel was amazing, and they gave me an oceanfront room looking directly at Waikiki Beach. It was by far the nicest hotel I had stayed in to date, and the service was exceptional. I had finally made it to Hawaii, and so far, it was everything that the pictures represented.

My second evening there, I had reservations for the Al Harrington Polynesian dinner show. Because I had booked the agent discount for the show, I was considered a special guest and at the start of the show, I found out just what that meant. Al Harrington is an actor/singer who played the role of **Ben Kokua**, a detective on the original Hawaii Five O. After the show ended, he became a performer in Waikiki with his singing, and Polynesian dancers. At the start of the show, he came out and said he had some special guests at the show he wanted to introduce. My name was one of the ones that was called, and he had me stand up and wave to the crowd, thanked me for supporting the show, and even made a joke about Pittsburgh. It was totally unexpected on my part, and just another example of how I was able to experience something that I might not normally have been able to. The show was great but by the time I had made it back to Hawaii on a later trip, Harrington had retired from performing, so I did not get to see the show again.

The Al Harrington Show July 1987

No trip to Hawaii would have been complete without attending some sort of luau, and there is no shortage of options available. The one that was recommended to me was Germaine's Luau, and it did not disappoint. It by far is one of the biggest as there were hundreds of people who were there the night I went, but it never seemed too crowded. The traditional items were present, the emu in the ground, all the other fixins, and unlimited Mai Tai's and Blue Hawaii's. There was also a Polynesian show, but it was pretty much what I had seen the night before.

My final day in Hawaii was the one true sightseeing day I had planned, which meant a trip to **Pearl Harbor, Punchbowl Cemetery,** and the **Iolani Palace.** Pearl Harbor was every bit surreal, and amazing all at the same time. Seeing the USS Arizona still laying at the bottom of the harbor, and learning the full story made it very hard not to be moved by the whole thing. The other historical sights rounded out the rest of the day, but it was now time to head back to the hotel and head to the airport for my overnight flight back to the mainland. I was very glad I worked this trip into my career search, because as I would find out, my professional life would take a priority over my travel pursuits for the coming months.

Turks & Caicos
MY FIRST CLUB MED, NOVEMBER 1991

During my first couple of years working on the leisure travel side of the business, one thing I always looked forward to, was when the new **Club Med** brochure would come out. The pictures from all the different clubs around the world looked amazing, and I thought about which ones I would like to go to. For those who don't know, Club Med is a member's only program that you can join, and then travel to any one of their clubs around the world. They encompass a mostly all-inclusive concept, where everything is included, except unlimited bar drinks. They do offer wine and beer with meals, but all bar drinks must be purchased as additional items. One of the clubs that caught my eye was called **Turquoise,** located in the **Grand Turk Islands.** Both the water and the beach looked amazing, and the club offered a host of water sports that are included in the price. Most of your money goes into the activities, as

Club Med accommodations are basic, and not really posh. The big question for me was how I would check this place out, and if I did, would it be as spectacular as the brochure made it out to be?

The answer seemed to be like many others, let's have some kind of contest. At this time of my career, I was working with American Express, and Club Med was a preferred vendor of Amex. Once a year they would do a big Club Med promotion, where the office would decorate, and the Sales staff would really try and promote Club Med vacations. There would be a grand prize for the best decorated office, and in addition, there would be a Club Med mystery shopper, who would call Amex offices pretending to be a traveler looking for vacation recommendations. As it turned out, our office won the best decorated contest, and a cash prize, and yours truly, received a mystery shopper call. I did not know who it was, but it just so happened, I recommended Club Med to the person. Afterward, the shopper told my manager that I was so enthusiastic about selling the concept, he felt bad he wasn't actually a real traveler. Because of that performance, I won a trip for two to any Club Med in the North America region. A few months later, we were packing our bags for the Grand Turk Islands.

Getting to the Grand Turk Islands from California was not the easiest of trips, mainly because of the limited number of air lines that flew into the Island of **Providenciales.** We ended taking American Airlines to Miami, then flying **Pan Am** (remember them?) from Miami down to Provo. I was excited for my wife because she is much more of a beach type person, as opposed to sightseeing. Up till now, most of the trips we had taken as husband and wife, were sightseeing in nature. This trip was just for us to relax for one week in the Caribbean sun, and not have the worry of budgeting money, or looking for activities.

When we arrived at the Club, we were greeted with champagne, and I had a slight flashback as to whether they might hand us sheets for a toga party, but that was not to be. Our first night at dinner, they sat us at a table with other guests, and the introductions started. As humans tend to be creatures of habit, Mary and I, along with our newfound friends hung out together pretty much for the whole week. Two members of our group were young ladies from the New York City area, Moira and Alyssa. They were a bit closer in age to us, so we had an instant click. As the week went on, so did the fun, although the weather was not cooperating as we had several days of all-day rain, but you really try to make the best of it. There was still time to get in water skiing, snorkeling and for a first time for me, parasailing.

Also, a first for me was this very subtle attraction I had for Moira, which was the first time I had that feeling since being married a year and a half earlier. I suppose when you are still young, you sometimes wonder if you were ready to settle down, and in my case, I believe I was, but I always had an appreciation for beauty and kindness. Moira was never aware of this and is learning about it for the first time as she reads this. As I have often done on many of my trips, I remain in contact with friends that I have met. I am pleased to say that Moira and I are still friends, and I have had the opportunity on some visits to New York, to meet up with her for dinner. She is a very successful Marketing Executive, and I must say I am envious of her, because she has traveled to more exotic places than I have. Everyone mentions how important it is to have friendships in your life, but I have the added blessing of having friends around the world, and Moira is one of them. A friendship that has endured from a long distance for almost 27 years.

Saying good bye to new friends after a week is hard, but as I have found out in some cases, it is more like "see ya later". The return home was uneventful, but we were lucky because 3 days after we returned home, Pan Am ceased operations, and grounded all planes. It would have been a nightmare trying to get off that island, had one of the only airlines flying in, stopped operations while we were there. We had a great time, but somewhat surprising is that I have never gone back to another Club Med since that trip. Of course, the life changes that were soon to come, had a big role in that.

The whitest sand and bluest water in the Grand Turk Islands, November 1991

Hawaii (Part Deux)
OCTOBER 1998

Quite a few years had passed since we had last taken what I would consider to be a significant vacation. We had done some week end trips here and there, but nothing really that I would classify as a full-fledged vacation, and there was a good reason for that. During this period of time, Mary and I had started a family, and were working towards purchasing our home. With these priorities in place, the luxury of time and money was something that had become secondary. In 1993, our daughter Alissa was born, and in

38

1996 Katie came along to join our family. Two young ladies provided enough work to keep us busy, and now because of a growing family, we needed a bigger home. Then in 1998, we bought a new home where our kids could grow, and no longer have to share a bedroom. My career had also grown in that I had made the transition from selling travel, to working the technology side of things that helped sell travel. In addition to that, I had been working on my formal education, and completed a bachelor's and master's degree from **California State University Fullerton.** Once I graduated in 1998, and we were moved into our new home, we thought it might be a good time to celebrate our achievements and take what would be our first true family vacation.

Even though I was working in a different area of the travel industry now, I was still able to take advantage of some of the perks. Since I now did a lot of business travel, free airline tickets gave way to free frequent flyer tickets. I was also still able to retain my travel credentials that allowed for hotel and rental car discounts. The choice where to go wasn't hard when both Hyatt and Sheraton had a "3 nights free" promotion, and I had enough miles on Delta to get us to Hawaii. This trip would be much different for me than the last time I traveled there 10 years earlier. The late-night luau gave way to the Honolulu Zoo, and the afternoon Mai Tai specials at Dukes were traded in for building sand castles on Waikiki Beach. This was what a family vacation was supposed to be, doing things with your children to remember. Although as Alissa was only 5, and Katie not quite 3 yet, I'm not really sure how much they took in. The other part that was new for us, was the lack of romantic dinners, or even any dinner for that matter. We had exhausted the girls the first day so much that their afternoon nap turned into an all-night event. This is where I got creative, and simply went and got Mary and I some take-out food, with a stop at the **ABC store** for a bottle of champagne, and we had dinner on our lanai facing the ocean.

Trying to keep two little girls entertained is not that difficult when you have miles of beach, and a big ocean to keep them occupied. We also took them to the Waikiki Aquarium to let them see what was in the ocean up close. Letting them have the experience of an airplane ride, staying in hotels, and even getting to hold a beautiful Cockatoo bird for a family picture was something that made them smile. I experienced an even bigger smile on our final day there, because of the actions of my little Katie Girl. We had scheduled an overnight flight on our return to allow more time there. I also had reserved a day room at the hotel, so we could get cleaned up and rested before we left. All day on the beach was not enough for Katie, as she heard my voice call her and tell her it was time to come in, so we could get ready to leave. As if on que, she heard me say "leave", and took off running down the beach, and would not stop. She ran from the beach at the **Sheraton Surf Rider,** down past the **Sheraton Waikiki** beach until I finally caught up with her. Part of me was laughing, and part of me was frustrated, because she knew what she was doing. As a stroke of good fortune had it, we caught that moment on video, and have laughed at it each time watching it.

Our first family trip to Hawaii, October 1998

The most magical thing about this trip was the sense of pride I felt at being able to give my family a nice vacation, and seeing the smiles on my children's faces, and the excitement they demonstrated. The success of this trip had me immediately planning something even more grand in the future. I knew they were only going to grow up, and learn to appreciate the concept of traveling more, but as I had learned on many trips before, you never know what life will put into your path.

Disneyworld/Cruise
JUNE 2004

Again, I would find myself in a long stretch of time between family vacations, as a few new events unfolded from the original plan. The 1st was that we welcomed our 3rd daughter Megan into our family in March of 2000. Megan was not a surprise; however, my wife and I talked about a 3rd child, and if we did have one, we wanted it to be by the turn of the millennia, or we would be happy with our 2 daughters. We were free from diapers, and just at a point were behavior was not an issue when doing things. Then we found out about Megan and realized that we had one more blessing added to our lives. In addition to that, Mary had decided that she wanted to return to school, and obtain her master's degree in teaching, so she started that program. As she was set for graduation in June of 2004, we thought it would be a good time for a family vacation, as a way to celebrate her achievement.

It seems inevitable that at least one family vacation has to focus around the Magic Kingdom of Disneyworld in Florida. It might seem strange that someone who lives 35 miles from Disneyland in California, would feel the need to head to the Florida park, but everyone goes there. Let me preface by saying I am not a fan of Disney at all, and patronizing them in any way, goes against my grain. I do understand however, they appeal to kids and to some adults, and this trip was about the kids. My personal feeling is that they gouge the consumer anyway they can, and the overall value for the money spent is a poor buy. My opinion on this is probably in the minority, but it is one that I have always held. My thought on this trip was, we were going to have to take it at some point, so let's get it over with. Megan was now 4 years old, and all 3 girls would have memories of this, so now was as good a time as any to go.

As a way for Mary and I to get some enjoyment out of this trip as well, we decided to toss in a 3 day cruise to the Bahamas on Carnival to make it a complete trip. We started the trip by flying to Ft Lauderdale with free tickets on Delta Airlines, as I had enough miles to cash in for the 5 of us. While we had not taken a family vacation for several years, I was still traveling extensively for business, and racking up miles on Delta. Thus, when it was time to head to Florida, Delta was king. We spent an extra night in South Beach before the cruise and stayed at the Fountain Blue Hilton right on the beach. One thing I remember so clearly was the first time the kids went swimming in the Atlantic Ocean, they were surprised how warm the water was. The Pacific Ocean in California is quite cold all year round, and they could not believe the difference in water temperature. The next day was the first day of our cruise, so we headed to the Port of Miami, and boarded the ship.

As always seems to be the case when traveling with children, one of them gets sick. This trip it was Megan who developed a cold and visited the ships doctor. We managed to get a snorkeling trip in when we reached Nassau, and Katie got her hair braided like the island girls had theirs done. The ship provided a lot of activities for the kids, and it was a great experience for their first cruise. It was a concept that we would revisit in the future when another family vacation would come about.

After the 4-day cruise ended, it was time for us to drive from Miami up to Orlando and start the Disneyworld portion of this trip. We had 4 days' worth of Disney planned, and purchased a 4-day passport for each of us. Hilton Hotel points covered 4 nights of the Homewood Suites, which had a 2-room suite that is ideal for families. Our 1st Disney Park would be the **MGM Studios**, and like all of the Disney parks it was packed to the rafters. The lines were incredibly long, the sun was incredibly hot, and the kids were less than happy, and we won't mention how I was. Especially hard on them was the humidity of Florida, as this was something they were not used to at all. A long day led us back to the hotel for a good rest because the next day was The **Magic Kingdom**. Disneyworld surprisingly went off without any hitches, but the walking did take its toll on the girls. After this day, I was starting to perk up because the next day was for me, even though I said it was for the girls.

There isn't anything better on a hot day than a cool water park, and that's what was waiting for us at **Blizzard Beach.** A water park themed like a ski resort, I was pretty jazzed about this day, and we were one of the first people there at the opening. Soon after, we were no longer the first people there as thousands more showed up, and the park reached capacity for the day. To make matters worse, Alissa and Katie got separated from me, and I lost them. Megan and mom sat this day out, and chose the hotel pool, and it is a good thing they did. For over 2 hours I searched for Alissa and Katie, whose idea to search for me included riding the slides as they looked. I finally had to file a missing child report with the park police and have them help me locate them. While they enjoyed their day, I ended up on about 4 slides, and a sunburn to show for it.

Our final day was spent at the Disney Animal Kingdom, and aside from seeing some animals, and the famed **Tree of Life,** it wasn't that thrilling of a day. I could tell that everyone was tired and ready to head home, a byproduct of a lot of traveling, and the hot Florida weather. I also could tell that the girls had a great time, as I could see it in their faces, as they were now starting to really understand the concept of family vacations. I did have some concerns that they might be getting the notion that all family vacations are this elaborate. I wanted to impress upon them regardless of what you did, as long as you were with your family, it could be a vacation. We just happened to be fortunate that the travel perks I had accumulated allowed us to take these types of adventures. As much as I really was not looking forward to taking this trip, I realized how important it was to put aside my own personal feelings, for the sake of my children.

Mexico
OCTOBER 2006

It was only 2 years removed from our Disney extravaganza, but much had changed in my life. At this point in my career things started to really take off, (no pun intended). My leap into the Sales Department saw my business travel more than double, as I was involved in almost every sales engagement we were a part of. I had also switched over from Delta to United Airlines as my preferred airline, primarily because Delta revised their loyalty program, and made it increasingly harder to obtain the perks I was so used to. It was also around this time that initial cracks started to appear in my married life, as we were 16 years into wedded bliss.

Because I had done so much traveling the previous 3 years, I was able to accumulate 750,000 miles within my United account, and I felt it was time to use them. There are always rumblings that the airlines want to do away with loyalty programs, and I had too much invested to lose that many miles. Based on available miles, we talked about taking the girls to Australia, and giving them a real experience. As I mentioned in the Australia chapter, I thought it was a nice place, but it didn't thrill me enough to place it on my list of "must returns". Still, it offered free accommodations, and with free airline tickets, it was a hard deal to pass up.

Based on my calculations, I had enough miles to cash in 5 Business Class tickets to Australia, so we decided to go for it. Then the concern I had about my kids becoming a bit spoiled with the travel perks was realized. When I checked with United Airlines about availability, they were very direct in telling me that they would NEVER clear 5 free Business Class seats on the same plane at any time. They said they would clear 5 Coach Class seats, but not Business, or I could break out the number of Business seats on two different flights. The second option was not feasible at all, but I left it open for discussion. I explained the situation to the family and told them our best option was to go Coach Class for the convenience. To my surprise, my daughters and my wife stated they did not want to go Coach and would prefer not to go to Australia. That decision left me a bit flustered as my children had never flown anything but Coach, so why the sudden insistence on a superior flight? Mary and I had made that trip in Coach before, and indeed it is a long flight, but to scrap the whole trip altogether?

Since Australia was out, I made an executive decision and cashed in the miles for 2 cabins on a 7-day cruise to Mexico. Since our first family cruise was a success, I thought we could have a relaxing time for a full week visiting several areas of Mexico. The cruise would take us to Puerto Vallarta, Mazatlán and Cabo San Lucas, as well as a couple of days at sea. The nice thing about a cruise is that there is a lot for every one of every age to do, so I thought Mary and I would have an opportunity to relax, and work on things for ourselves. But ultimately when you have 3 kids, there will be times when everyone doesn't agree on what to do. This led to a lot of bickering on all fronts and increased the stress level between Mary and myself.

So, making the best of everything we arrived into Puerto Vallarta where we arranged for the girls to have a dolphin adventure, and actually swim with them. It was a great experience for them topped off with a custom photo of each and their dolphin. The girls also got to experience some of the fine dining elements of a cruise, as Katie wanted to try lobster, and all the girls joined their dad for midnight cheeseburgers on more than one occasion.

Next it was on to Mazatlán for what was the hottest day of the entire trip, spurred on with high humidity. A local tour of the city was somewhat interesting for me, but the girls found it less so. Lunch at a little cantina made for a nice break and the girls had their picture taken with a Gila Monster, who apparently is a regular. Our final stop was the highlight for me as we pulled into Cabo San Lucas. Being a part of Baja California, the weather in Cabo was much drier than Mazatlán, and therefore was not as uncomfortable. This was where we really got to partake in the fun of the water. Alissa wanted to try parasailing for the first time, and I had my sights set on renting a wave runner for a couple of hours and cruising around the bay. Each of the girls would take turns going out with me hitting the waves at a fun speed, and then refusing to let the other have a turn. It was the kind of day I had been hoping for when we sat down and planned out this trip.

As the ship pulled into Long Beach Harbor, it had been a fun time for the girls, but ultimately, I don't think Mary and I achieved our goal of what the vacation should have been. It was a theme that would be revisited on future trips, and ultimately take me into significant life changes.

Wave running in Cabo San Lucas , October 2006

Grand Canyon
JUNE 2007

Up till now, most everything I have written about has seemed like exotic type trips or vacations. But a life changing trip, does not need to encompass the fancy air, hotel, car concept. There are many wonderful places within our country that can be experienced, and many you are able to drive to. Since we upgraded to our dream home in 2005, and most of my travel perks were used on the cruise the year before, we thought about doing something a little lower key, but still a new adventure for us.

As long as I have lived in California, I had never made it to the Grand Canyon. Though not that far away, it just never seemed like a trip that I wanted to take. But in 2007, looking to do something fun, but stretch the ever-shrinking dollar a little further, we decided to take a trip to the Grand Canyon, and also include some water fun along the Colorado River. This was really our first significant driving type vacation, complete with luggage on the roof, and minivan packed with sleeping blankets and snack food. Yes, we were driving to the Grand Canyon, but somehow, I sensed this would not be the adventure that the **Brady Bunch** took when they went.

In fact, when the trip started, it had the makings of **National Lampoons Vacation,** as opposed to the Brady Bunch. As we rolled down Interstate 40 between **Barstow and Needles**, one of the suitcases from the roof came loose and fell onto the freeway. Fortunately, as it was still early in the day, the traffic was light, and we were able to retrieve it without incident. The drive between those two cities is 100 + miles of pretty much nothing, and hot as blazes in the Summer time. Our destination for the day was **Flagstaff Arizona,** and as we passed into Arizona, we started to climb into the mountains. Flagstaff sits at approximately 7000 feet above sea level, so the temperature there, even in Summer is considerably cooler. It was here we would spend our first night, and then start an early day taking the train to the Grand Canyon the next morning.

Rather than drive to the South Rim of the Canyon, we opted to take a train into the Canyon area from **Williams, AZ** on the **Grand Canyon Express**. I had heard that parking a car at the Canyon was a nightmare, and taking the train was a popular way to go and save the driving hassle. The train left around 8am, so we had to be at the station no later than 7:30am. When we arrived at the station, there was a Western style show underway, including cowboys, and fake fight scenes. Then our brush with celebrity happened as the girls kept saying to me, "Dad it's them, it's them". I looked over and said "who", and they responded, "the people from TV". As they pointed, my politically incorrect voice took over and said, "you mean the midgets". Right away I was corrected, and they said "Dad, they are little people". I still didn't know who they were, but apparently there was a TV show called **Little People, Big World,** and this family was taking the train to the Canyon like us, as they were filming an episode.

Celebrities aside, we arrived at the Canyon about 2 hours later, and were able to walk to the viewing areas to see one of the 7 natural wonders of the world. The Grand Canyon from any angle is an impressive sight, and I was glad we were able to share that with the girls. We were able to have someone take a great family picture with the canyon as the background, and a friendly little squirrel photo bombed us at the perfect moment. Seeing the Grand Canyon was a wonderful experience, and while I wasn't paying attention to this little piece of trivia, it was the 3rd of the 7 Natural Wonders of the World I had witnessed.

Now it was time for some more fun, but no vacation would be complete without a trip to the doctor with a sick child. So before leaving Flagstaff, a trip to Urgent Care with Megan was on the list. Being more than a little annoyed, my frustration soon turned to concern as my little girl was diagnosed with a slight case of pneumonia. Interestingly enough, they told us that she could go enjoy the water, but to not overdo it, and to take her medicine. So, we continued on with the plan, and headed to Laughlin, NV for a couple of days of wave running. When we left Flagstaff it was 75 degrees, and upon arrival in Laughlin, it was 104. This area of the Colorado River is extremely hot in the Summer time, and generally I am good for just a couple of days in this high heat. We rented a wave runner, and took turns going up and down the river at pretty good speeds. Mary would generally sit these types of adventures out, as laying on the beach was more her thing. I always felt that she missed out on the fun of growing up with the girls, but it was clearly her choice. After a couple of days on the river, we headed home with another experience in our hip pockets, and one that was very different from some of the things we had done in the past. Little did I know that I was only a month away from a fundamental life change that would start my life on an entirely different direction and transform the person I would become.

Lake Tahoe
FEBRUARY 2008

With the kids getting older I started to feel that there was something missing from the experiences we shared. We had never taken a Winter vacation, primarily because there is essentially no Winter where we live. But having grown up in a four-season climate, and being a Winter sports enthusiast, I wanted to have them experience skiing. Alissa had gone skiing a few times with me before, but Katie and Megan were much younger then. Now, everyone was at the age where they could try something new and be able to enjoy the experience. I really wanted to show my girls the fun that you can have when the weather is cold, and not just do the same things day after day. The local mountains in Southern California offer skiing, but the season is so limited based on warmer weather, it is often not worth the effort to try. So, I wanted to take the family somewhere where we could pretty much count on great Winter conditions.

Some of the best and most abundant skiing in the West can be found in **Lake Tahoe.** An area in the Sierra Nevada Mountains, Lake Tahoe is shared by two states. 2/3 of the lake resides in California, while the other 1/3 is in Nevada. The area has over 10 quality ski resorts, and some very famous ones, such as **Squaw Valley,** home to the 1960 Winter Olympics. Feeling pretty confident that we would have some good skiing conditions, since the area averages about 400 inches of snow a year, we decided to take our first Winter family vacation to Tahoe. Since this was a family trip, we decided to drive, even though it is a little over 500 miles, and would take a full day, it seemed a little more relaxing than dealing with plane tickets and luggage etc. To save a little time and enjoy optimal time there, we decided to drive through the night, for an early morning arrival there. That plan worked as we were fortunate the weather cooperated, and we did not run into any storms along the way.

Given this was our first real Winter vacation, I wanted to plan some things in addition to skiing, and on our first day there, I was able to arrange a sleigh ride in the area. It was a chance for the girls to experience a real horse drawn sleigh with a view of the lake. An added bonus was a chance for the whole family to have our picture taken with the horses, which are truly magnificent animals. With the help of Hilton Hotel points, I was able to confirm our stay at the Embassy Suites in South Lake Tahoe. This hotel is right on the shore of the lake and sits right at the base of the Heavenly Resort Gondola. Again, staying in a suite type hotel is perfect for families as there are two separate rooms, and plenty of space for you to stretch out. We needed that rest, because day two would be our first day skiing, and would make for a long day.

With so many resorts in the area, it was hard to choose which place to ski at. There are many factors to consider such as type of terrain based on skiing ability, size of crowds, and of course cost of lift tickets. Since 2 of my daughters would be taking skiing lessons for the first time, I wanted a place that offered a good program for them to learn, without breaking the bank. We chose **Mt. Rose** resort on the Nevada side, because of their first-time ski special, and because they offered a good mix of terrain for all levels of skier. I had skied Rose before, and really liked the layout so that each person can ski to his ability and meet at the bottom of the hill which was centrally located. This allowed me to hit some of the higher runs, and still meet Alissa on the way down in order for us to head back up again together.

By the afternoon hours, the girls and Mary, who had decided to take a refresher lesson, were completed with the lessons, and joined us on the main mountain. I was literally blown away at how fast Megan at 8 years old had taken to skiing so quickly. She was at the age where no fear entered into her mind and was skiing fearlessly down the beginning mountain. I stopped to take some video of my family all together enjoying something that I grew up doing, and with some regret in my mind about why it took so long for me to introduce them to this part of life. Adding to this great day was the fact that my Sister-in-Law Erin, and her family were able to join us as well. They live in the Tahoe area, and were able to take a day and spend it with us. That day started a trend where we would spend time with them skiing each year and enjoying a few days in Winter doing what families do. I grew up skiing back in Pennsylvania, but I never experienced the feeling of a true family vacation in the mountains until this trip.

Family skiing in Lake Tahoe, February 2008

Our second day of skiing had us trying a new resort in South Lake Tahoe, called Sierra at Tahoe. I have skied many different resorts in my life, but there have only been a few that I can honestly say I did not like, this was one of them. I had read that they were very family friendly with terrain that was suitable for beginners, but some runs for the advanced. What I found was terrain that while the vertical grade was mild, the runs were very narrow, and surround on both sides by trees, and closed off areas. This became a scary problem when Megan, who mind you is still learning, but showing no fear, picked up more speed than she realized, and could see an oncoming turn just ahead. She started to panic as I was next to her, and knew she could not make the turn, and would have ended up in the woods, possibly hurt. At that moment I told her to fall down to avoid crashing into the net, and for a split second she was afraid to do that, but I believe instinct took over, and she just dropped to the ground. She was fine, but I believed that moment brought the element of fear into her, and she was never the same skier after that. She became a much more cautious skier, and while you want to practice safety, being too cautious can inhibit your ability to improve your skiing.

On our final day, we took a break from skiing, and decided to do something a little more retro, which was inner tubing. I am all about having fun in the snow, whether on skis, ice skates or inner tubes, it is all fun to me. We were fortunate enough to have Erin, and my two nieces Emilee, and Michelle join us for this fun day. It was wonderful to watch my daughters, and their cousins spending time together, and doing it without the sun and sand of the beach. Yes, those are nice days as well, but it was very special for me to introduce my family to the wonders of Winter, something they had not been exposed to before. The success of this trip led to several additional ski trips in the subsequent years, and a new adventure for my girls.

Mexico
JULY 2016

There was a significant stretch of time between the last family vacation, and where it brought me to this point. It wasn't as if we did not do anything as a family, as we continued to take small ski trips, and other things as a family; however, this period of time was also one of the darkest of my life up to this point. During the stretch of 2010 to 2016, I had experienced life changing events, such as losing our home, the death of my best friend, the death of my brother and mother in a span of 60 days, a divorce, and a broken engagement. Given all of that, a vacation just wasn't at the top of my list, but it should have been. For a couple of years, the girls had mentioned that all we do is take Winter trips, and they were ready for something more along the lines of Summer. I gave it some thought, and they were right, so I made a commitment to take a family vacation with them during the Summer of 2016. This would be our first family vacation that did not include their mom, whom by now was my ex-wife.

The other part of this is my girls had become adults, and at what point do you move away from the traditional family vacation mode, and they start doing things on their own? I did want to do something special for them, but I also needed to establish some guidelines for my own life. I had felt like I was languishing in a rut, and not making any progress to move forward with my own life, so I had a very frank discussion with them. I wanted them to know that this would probably be the last family trip for a while, as I needed to find my purpose, and do some things that I wanted to do. The concept was foreign to me, but all of the experts I talked to said this was essential for me to start doing some things for myself.

I decided to take the girls on another cruise because I felt this would be a good way for everyone to do whatever they wanted to do. No one would be forced to do something they didn't want to do. I could sit on deck and write if I wanted to, and they had the run of the ship to do what they wanted to. They agreed with the plan, and I was even able to add a little extra to it. My buddy Jay owns a condo in **South Beach**, about 3 blocks to the water, and he allowed us to use it for four nights. That meant we got to experience 4 nights in the Miami Beach area and would also have a 4-night cruise visiting Cozumel Mexico. Our time in South Beach was relaxing and included meeting some old friends Tim & Tina for dinner. We also got to take a tour of the Everglades, and search for alligators which was my first time experiencing that.

Upon leaving South Beach, it was a quick jaunt to return the rental car and call an Uber for a ride to the Port of Miami. Here we boarded the ship, got squared away and headed for the first stop which would be **Key West** the next morning. The first night on the ship, all of us checked out the comedy show on board, and basically just explored the massive ship. The next morning in Key West, Katie chose to sleep in, while Alissa, Megan and myself went ashore. This was fine because I wanted everyone to do what they felt was a vacation to them. For Katie, that meant sleeping in and passing on Key West. The time in port was really only a couple of hours, so there wasn't a whole lot you could schedule. Alissa decided she wanted to look for some souvenirs, Megan decided to check out a museum, and I just wanted to find a fancy tropical drink. Soon we were back on the ship and heading for Cozumel with a scheduled arrival of 12 noon the next day.

For some unknown reason, we were behind schedule with our arrival into Cozumel by about 2 hours. This had me very concerned because I had booked a full afternoon tour that included dune buggy off roading, a Mayan Ruins visit, a snorkeling adventure, and lunch on the beach. I had told the girls in advance what my plans were, and if they wanted to join me let me know, and if not, they were free to do something else. All of them chose to go with me, which I was glad they did, because as it turned out, it was a lot of fun. I was afraid we would miss our tour time, but what I learned was your tour time starts when you show up. We got into our dune buggy, which was nothing more than metal pipes, and a gas-powered engine, but it was really cool looking. Our guide led the way as we were driving through Downtown Cozumel on our way to the Mayan Ruins. It took about 20 minutes to get there, and when we arrived, we had the opportunity to see remnants of a 2000-year-old civilization. I'm not really sure how interested the girls were in all of this, but it was something I wanted to do. Next it was on to the off roading adventure along the sea, and I made it a little more intense than it needed to be with my running through a huge mud puddle and crashing into a tree along the way.

Off-Roading in a dune buggy Cozumel, July 2016

On the way back to town, they had us stop at a little family owned tequila plant. These kind of shopping stops are not uncommon, but it did give the girls and I a chance to taste some local tequila. Megan being still under 18 was excited that they let her taste as well, and as a gesture, I did purchase a bottle to go. Our final stop was back at the club for a little snorkeling session where Megan got to try out her Go Pro type camera under water. As we got back to the pier, I had enough action for the day, and decided to head back onto the ship, but the girls wanted to shop for some gifts a little longer. It was nice to be able to let them do what they wanted, without it hindering what I wanted to do. They were lucky enough to find someone to take their picture from the dock at twilight with the ship lit up as the backdrop. That photo became my favorite of the entire trip, and really highlighted that my girls were no longer little kids but grown adults.

On the ship I was really excited that they had a formal dinner event, because it gave me a chance to break out the tux and allow them to where evening dresses. I know they were a little annoyed that I wanted to take so many pictures, but this was a pivotal moment in my life, where a new family dynamic took shape. Truth be told, I think they liked the formal pictures we had taken as well, because it let them see that they were now grown up as well and traveling their own path in life.

As our trip came to a close, I started to reflect on where things were headed for me in the future, and how I would walk that delicate line of protective Father, and letting my girls seek their own adventures. I have no doubt that each of them will achieve their goals, as I instilled in them to take advantage of travel, and experience the world, before it becomes too late. I know this trip will not be the last that we take together, but I know they are anxious to have their own experiences, that will probably differ from mine. I could not be more thankful, blessed, and proud of each of them, and the adventures that we have shared in this life.

My 3 young ladies Cozumel, July 2016

Moab, Utah
October 2019

Taking a work-related trip and extending it into something fun and adventurous was not really a new concept for me. I have often done fun things whenever I travel for work, if it is at all possible. What was different was having a work trip that was based solely on fun and team building. Most of the companies that I have worked for, felt that in order to get the maximum benefit out of employee meetings, they had to jam pack the agenda with nothing but work-related information, and disregard the concept of information overload. That however is not how the company that I work for now operates, and the validation of that was a Business Development meeting that we had in the city of Moab, in Southeastern Utah.

Moab is a small town, that is home to 2 of the 5 National parks that reside in Utah. It is a destination ripe for the outdoors type that extends beyond just hiking, and includes mountain biking and a variety of "Off Roading" adventures. It was some of these adventures that the meeting focused on, as well as true work items.

As many times as I have been to Utah, I have never been to this area, so this was a new adventure for me personally. Since I love the outdoors, I was really ready to make the most of this trip, and fit as much activity in as I could. That would require some early risings and intense hiking, as well as a willingness to get a bit dirty. With so much to do and only 4 days (2 work days & 2 weekend days) to do it,

I had to do my research to discover the things I truly wanted to experience.

Day 1 started with an optional morning hike to an area called **Corona Arch**. The geographic area we were at consists of thousands of sandstone arches, and the Corona Arch was just one of them. Starting at 6am, the day was still dark, so a group of 27 people started out equipped with headlamps in search of the Arch. After about 45 minutes of hiking, we arrived at the location just about daybreak. The Arch was a unique formation, and it leads the mind to wonder how on Earth did these natural formations get made. Having had an opportunity to stand underneath the structure, it really brought into focus just how large of a structure it was. A few pictures later, and a few silent pauses of awe, and we were heading back to the trailhead for a return to the hotel. A great start to our morning, that would soon give way to a real adventure in the afternoon.

After a couple of hours meeting as a group, it was lunch time, and also time to pick up our 4X4 vehicles for an Off Roading adventure. This would be something new for me, and also a chance to try out my new GoPro action camera. Our group had 9 vehicles, with 4 people per vehicle ready to hit the off-road trails of Moab. It was recommended before the trip that we have a pair of protective eyewear for this activity, and indeed I bought a pair specifically for this. I was told that the dust generated from the vehicles can be significant, so protective eye wear is a must. That was probably the biggest understatement I have heard in a long time. The dust was so severe that I really should have used a pair of ski goggles to protect my eyes. As anyone who wears contact lenses will tell you, even a speck of dust can cause a tremendous amount of havoc with your vision. So, for about 2 hours I battled the elements and my watering eyes, which is the main reason I chose not to drive one of the vehicles. Normally, I am all in for trying something new, but I knew driving one of these 4X4's was not in anyone's best interest. I must say that I was very content to be the passenger for this activity, as it allowed me to take some incredible videos of the fun and the scenery.

Off Roading Adventure, Moab, UT
October 2019

Day 2 again led off with an optional morning hike, this time to **Arches National Park** to see the world-famous **Delicate Arch**. This is the Arch that is seen in all Utah tourism collateral, license plates, and a plethora of photographic pictures. The monument draws a huge amount of amateur and professional photographers seeking their pinnacle sunrise or sunset photo of the Arch. The hike was about 1.5 miles in each direction, and was so worth experiencing, even as the morning temperatures were in the low 40's. One little disappointment in this hike was that the weather was cloudy, so there really was no sunrise to observe, but the arch was impressive nonetheless.

Our afternoon consisted of a couple of hours of meetings, and then it was time to wrap up the business part of this trip. Most of the attendees would be returning home that Friday afternoon, but a few of us were staying on. Since I had to be back in Salt Lake City for meetings the following Monday, I decided to make a weekend of it, even though I knew I would be on my own. This gave me the opportunity to define my own agenda, and focus on the things that I wanted to see, and at a pace that allowed for no real time table.

Having spent very little time in Arches National Park, I decided to go back there on Saturday, and really see the park in depth. I started out again really early, as I discovered that if you arrive at the park before 7:30am, you can avoid the $30.00 entrance fee, which I thought was a good trade off. My goal for this day was to see as many of the famous arches that I could, without having to do extensive interior hikes. I was really pleased that many of the famous monuments in the park, are within easy driving distance, and do not require extensive hikes in order to experience them. I got to witness **Balanced Rock, Double Arch, Turrent Arch, and Windows** Arches from one stop of the car, and about 1 mile of overall hiking. I then drove to another area of the park called **Devils Garden**, and hiked out to **Landscape Arch**. This Arch is unique in that it is the longest free-standing arch based on length, and is positioned on the side of a hill which makes for amazing views. Also to note, that due to natural erosion, there are parts of this arch that are very thin, and probably will not remain intact in the years to come. I am very glad that I got to witness this natural phenomenon, before it comes crumbling down.

Landscape Arch, Arches National Park, Moab, UT
October 2020

After spending about 4 hours inside the park, and having seen what I mapped out in my mind, I decided to head back to the hotel. Taking one last look to enjoy the scenery and marvel at the beauty of this area, I took my time exiting the park, and thinking about what I had seen, and when I might return. As I got back to the entrance, I was reminded how an area so pristine can quickly change, as the line of cars waiting to enter the park stretched well past a mile. I later learned that the park had actually reached capacity, and they stopped allowing people to enter, so it certainly paid in many ways to get up early, and go for it.

For my final day in the area, I learned about a state park called Dead Horse Point, and decided I wanted to see it. I had recognized the famous view from Dead Horse Point View, and subsequently learned that the horseshoe landscape of the Colorado River, is the most photographed scenic area in the world. This made it a "must see" for me, and so I set off on a Sunday morning to have my own view. Located about 30 miles west of Moab, the park is out in the middle of nowhere, and on this cold morning was sparsely populated, so I had the view pretty much to myself. It was everything that I had heard it to be, and gave me the opportunity to take some amazing photos, and videos of the area. 5 hours and numerous traffic jams later, I was back in Salt Lake City, and back into everyday reality.

Dead Horse Point, Moab, UT
October 2019

It's probably easy for me to sit back and ask myself why it took so long for me to experience this kind of an adventure. The truth is that I don't believe I would have appreciated this type of beauty during my younger days. It's only as I reflect in later years, that the appreciation for these natural elements, and the understanding that true adventure comes from the heart, not the location, has taken hold in my life.

Alaska
The 47th State
DECEMBER 2019, JANUARY 2020

Now before everyone gets excited and thinks they have discovered a critical error in my title, let me say that I am well aware that Alaska is this country's 49th state. It is however the 47th state that I have visited in a subtle, but very achievable goal of seeing all 50 states in America. I often see in social media outlets where people ask others to identify how many states they have visited, and airplane-change connections often count in that calculation. Well I am happy to say that I do not count airplane connections, and that I have visited 47 states at some level. Alaska was my 47th state and only the 3 M's of **Montana, Maine & Mississippi** have alluded me. I have mapped out a strategy to check Montana off my list in 2020, since my work headquarters is well within driving distance.

The larger question of why Alaska, especially at one of the coldest times of the year is complex, yet simple at the same time. First and foremost, I have always wanted to see the Aurora Borealis, the formal name for what we call The Northern Lights. The fact that The Northern Lights is one of the 7 Natural Wonders of the World, would allow me to check that detail off of my bucket list. If successful, the Aurora would be the 5th Natural Wonder that I have seen with my own eyes, leaving only 2 more, and making this goal that much more attainable. The other part of the Aurora is that it can only be seen in the dark, and therefore I chose a time where it is dark about 21 hours of the day.

There are other factors that are necessary to be able to witness this phenomenon, and I will talk to those in a little bit, but this is how I got to my 47th state.

Not long after I returned from my adventure in Russia, I received an email from American Airlines that reminded me of airline miles that I had and would soon be expiring. Since I started my new job in 2019, my travel patterns changed, and I started flying on Delta Airlines again, so my American miles had become somewhat dormant. I had just enough miles that I could cash in for 1 domestic ticket including Alaska. The Holiday Season was just around the corner, and I was looking at another year of the bachelor lifestyle, during a time of love and family. Like many people, I have a few scars in life that are tied to the Christmas Season, and sometimes I find myself struggling to get through it, wearing a forced happy face. I had often thought of taking a trip during Christmas as a way of avoiding the season, but never pulled the trigger on taking a Holiday adventure. The idea came to me about doing something epic for New Year's Eve. I think over the last generation or so, New Year's Eve has really lost its luster as something special. Given all the focus on drinking and driving (rightfully so), it seems most people would rather stay home. I have stayed home far too many times, and the idea came to me of doing something unique to ring in a new decade.

What if while most everyone watched a famous ball drop from the sky in New York City, I could witness the most amazing natural light show in the sky? Thus, the concept of seeing the Aurora on New Year's Eve was born, and now I had to make it happen. The closest place I could get to where I had a chance to view it was going to be Alaska, and since I had a free airline ticket that could get me up there, it became a no-brainer.

One of the most optimal places to view the Aurora, but still have some semblance of civilization was **Fairbanks Alaska**. The second largest city in the state is about 400 miles further North than the more popular **Anchorag**e, and much more rugged in wilderness. Much of my research had said this was the place in Alaska to see The Northern Lights, so that was good enough for me.

Always going a little extra has been how I have lived my adventures, and this was going to be no different. With that in mind, I looked at a couple of other "first time" things I could do, and mapped out a 5-day New Year's adventure. Getting as much activity into as little time as possible has served me well in other endeavors, so I kept that strategy intact, and set off on what I called my Arctic Adventure.

I set out for an Alaskan arrival on December 31st in order to keep with my New Year's Eve plan. I got into character right away as I was able to use my American miles flying on **Alaska Airlines**. I arrived into Fairbanks around 3am on the 31st, and since I had quite a few hours before I could check into my hotel room, I picked up my rental car, took a deep breath of the -1-degree temperature, and headed for the North Pole. North Pole Alaska that is, but still home to Santa Claus, and a town that is decorated for Christmas 365 days a year. The town of 3000 people plays to the whole Christmas concept, and indeed I got to speak directly to Santa Claus, who gave me some valuable tips on where to look for the Northern Lights later that night. After my visit to North Pole, it was time to head back and check into the hotel, and try to grab a little bit of rest, since light chasing is often done late at night for optimal viewing.

The Aurora Borealis

It was New Year's Eve, and the day that I would witness another Natural Wonder of the World, or at least I hoped I would. The Aurora is a natural phenomenon, and there is no guarantee that you will be able to see them on any given day. In addition to the conditions being dark, one other element I gave no consideration to was that it also needs to be clear conditions. The Aurora is not visible in cloudy conditions, and as I learned, Fairbanks is often cloudy during the Winter Solstice time. I found a local ski resort, that moonlights at night as an Aurora viewing station. Because the resort is out of town, it avoids much of the ambient light of the city, allowing for better viewing. There were probably about 40 people or so at this location waiting patiently for Aurora to appear. Like all of my trips, I managed to meet a couple from New York, who came to Alaska with the same goal as me, to chase the lights. They went by the names of Nub, and Zie, and while that is not their real names, I must say they had very Alaskan sounding names. I have mentioned throughout this book that the relationships I have formed from my travels can never be duplicated, and this was no exception.

As we found ourselves warming up in the lodge, the host came running in at 11:15pm, and proudly announced that "She is out". Aurora is referred to as a "she" in Alaska, and she finally made her appearance. We ran out to see what looked like a multi-colored rainbow in the sky. Only this rainbow was appearing on a pitch black, dry night. Over the course of 1 hour, the lights in the sky changed shape, dissipated at some points and offered a hint of color change. We watched in amazement for about 2 hours before the -6-degree weather started to take its toll on me. That, along with the fact that I had been awake for about 22 hours had me calling it a night at 2:00am. I also knew that I had another epic adventure planned for New Year's Day, one that would probably not be on most people's bucket list, and I needed at least a few hours of rest, before I set off on this new quest. One more item I got to check off on that bucket list, but more adventures to come.

Aurora Borealis, Fairbanks Alaska
December 31st, 2019

Adventure travel for me has always been about doing things that most other people don't do. Many people though travel to Alaska, so I wasn't really breaking any new ground here. However, when I learned that less than 1% of the people who visit Alaska, travel as far North as the Arctic Circle, I knew this was perfect for me. It seemed strange to me that very few people would make the trip, because the Arctic Circle starts at just 200 miles North of Fairbanks. That in terms of distance was not all that far, so it now became a quest for me to cross the Arctic Circle. As with any type of research, you are bound to learn new things, and what I learned was that a trip to the Circle was not something to be undertaken by an amateur, especially during Winter time. Rental car companies do not allow you take their vehicles on the highway that goes to the Circle, so you are pretty much relegated to booking a professional tour. So, I did just that, booked a 14-hour tour to the Arctic Circle on New Year's Day.

The tour included a few other sightseeing elements like the Yukon River, and several viewing positions for the Trans-Alaska Pipeline. Considered an engineering marvel, the Alaskan Pipeline stretches for 800 miles carrying oil from the Prudhoe Bay oil fields in the Arctic Ocean, all the way to Valdez Alaska in the Southern part of the state. Valdez is where ships filled with oil, depart to provide the rest of America with oil. On the surface, just a big long pipeline might not seem all that interesting, but when you think about the sheer distance that it runs, and what it accomplishes, you tend to gain a fresh perspective on how important it really is. It was now time to enter the Dalton Highway. This is the road that was created from Fairbanks to Prudhoe Bay specifically to build the Pipeline. It is a winding, unpaved road that goes on for 400 miles and has been made famous by a reality television show on the History Channel called Ice Road Truckers.

By the time we arrived at the Yukon River, it had been dark for some time, and we really didn't see much because of that. They say it is "mighty", and I guess you and I will just have to take that on faith, unless we return during a time when there is more daylight. From the Yukon River, I was still 2 hours away from the actual Circle, so we pushed North. Finally, at about 8:10pm, some 5 hours after we departed, we reached the Latitude lines of 66:33 signifying that we had crossed into the Arctic Circle. There was no fan fare waiting for us, no parade, just a sign welcoming us to this monumental place. Traveling all that way, over a very bumpy road, to spend 30 minutes in front of a wooden sign in -8-degree temperatures may not sound thrilling to most people, but I had now traveled into the Arctic Circle, and I became the 1%.

Euphoria became reality very soon as I realized that I was now going to have to head back and endure everything that I did on the way up. In the end, that didn't matter, I accomplished another first, and could just enjoy the ride back keeping a keen eye out for possibly an Aurora sighting. Arriving back into Fairbanks around 2:30am, with no luck on the Aurora as it was snowing quite a bit, I took stock of what I had just done, and knowing that a small percentage of people do what I did, I felt a sense of pride. I also knew that I had another first on my list of adventures coming up in just about 8 hours, so a nice rest was in order.

Arctic Circle Monument
January 1st, 2020

My passion for Winter sports is well documented, and I have done a lot of outdoor activities in Winter, but there is one activity that is extremely popular in Alaska that I have never done, and it was time to rectify that situation. Now you might not classify Dog Mushing as a Winter sport, but in Alaska it very much is. I am not aware of any other state in America where you can take a sled dog tour, so that meant I had to include this as part of the fun. Because this is a popular sport in Alaska, there are a fair number of vendors who offer these adventures. I scoured the Internet, compared different types of programs, and made a few phone calls. One place I called was Frisky Pups Sled Dog Tours because I liked what I read on their web site. In addition to price being a consideration, the other was duration of the tour. This is an outdoor event folks, and while I couldn't predict the weather, I had a pretty good idea it would be cold. With that in mind, a half day tour was not what I was looking for, but something along the lines of 1 hour.

I called and had to leave a voice mail, which is generally not a great sign of things, however I received a call back later that day from Bill, owner of Frisky Pups. He went over the details of the ride with me, and what to expect. I decided to go with them not just because of price and duration, but also because he asked about the details of my trip to Alaska. I told him everything I had planned, and he gave me some valuable tips on what to do, and what not to do. His interest in ensuring I had a great trip to Alaska, and not just a 1-hour dog sled ride, was the ultimate factor in my decision to choose his service.

So, on the morning of the ride, the thermometer was registering a brisk -11-degrees at 10:00am, and I started my 35-minute drive to the **Two Rivers** area of Fairbanks. When I arrived at the location, I was met by Bill and his wife Sandy, owners of Frisky Pups. Wearing my full-on skiing apparel, they proceeded to smile, and tell me "let's get you dressed properly". I guess I was not setup to be sufficiently warm for this ride, so they outfitted me with some heavy clothing, and as I would soon find out, I am so glad they did. Once dressed, we walked down to the kennel area where all the dogs are located. At first glance, I, like probably many others who are not knowledgeable enough about this, instantly thought that those poor dogs were freezing in this weather. They just looked so cold, and Huskies are not overly big, so it looks on the surface that this life is a cruel one. I soon learned that these dogs are bred for this exact thing, and there is more of a danger of them being too warm, rather than too cold.

Once the dogs were all linked up, we darted out of the gate and on to the trail. At last, I was a true **Yukon Corne-**

lius, and I even yelled out a **"mush"** a time or two. Funny thing is that Dog Mushers do not use the phrase "mush", that is a stereo type, but hey I am living a childhood fantasy, so let me have this. The reality is that when Mushers want the dogs to go faster, they say "hut, hut". A ride through the forest and across bridges eventually brought me to a frozen lake, where we stopped to admire the thousands of snow-covered trees. We had been mushing for about 45 minutes in -12-degree temperatures, and while most of my body was quite warm, my forhead and nose told me "don't push it". We headed back along much of the same trail, and I got to watch the dogs feast on some raw meat for a job well done. I pulled out from Frisky Pups with another new adventure under my belt, and a greater appreciation for just how animals adapt to their natural surroundings.

Alaskan Dog Mushing
January 2020

On my final day in Alaska, I had an open itinerary, with no real set plans. My flight did not leave until 8:30pm, so I had the time to do something, but what? The weather had turned significantly colder as the 10:00am temperature was a "mild" -20-degrees. I knew that I could not endure anything that required long periods outside, or could I? I had read about a thermal hot springs resort only 60 miles outside of Fairbanks, and immediately thought, "how cool would it be to go swimming outside, on a -20-degree day". Before I could say "Baby it's cold outside". I was on my way to **Chena Hot Springs Resort.** Along the way, I reached a great view point when the sun was rising, so I decided to stop and take a few photos. Granted the sun was rising at 11:02am, but it was still impressive regardless of the time. Not long after that I arrived at the resort, and immediately asked myself the question "do you really want to do this", and of course the answer was "first time, you bet".

I headed toward reception to buy my day pass, grab a towel and head into the locker room. Soon I was in my swimming suit, and headed toward the lake as calmly as if this was **Lake Havasu AZ**. The concrete walkway into the lake is heated, but make no mistake, it is bitter cold outside, and the key is to walk right into the naturally heated hot spring lake, and avoid that cold. The lake itself is around 104 degrees heated by the volcanic activity in the area, but hey, I was swimming in -20-degree weather. One of the trendy things to do is to wet your hair, and watch how quickly it freezes. Who am I to buck tradition, so I dunked my head and let it rip. Within two minutes, I had frosted hair, to the level that I paid a hairstylist $150.00 to do it. Because of the geothermal properties, the lake had a pungent sulfur smell, but who cares, I was swimming in Alaska, in the middle of Winter, what's a little smell. I spent a couple of hours in the lake, as well as checking out the standard outdoor jacuzzi, respecting the balance of nature where it was incredibly cold out, yet spending an afternoon in the water, scantily dressed.

Chena Hot Springs Resort (-20-Degrees)
January 2020

On my drive back to the airport, I had time to reflect on the natural elements that this trip brought to me. I was able to witness the natural phenomenon of the Aurora Borealis, the natural location of the Arctic Circle, the natural adaptation of Alaskan Huskie dogs, and the natural process whereby water is heated to a certain temperature, despite the frigid conditions surrounding it. Of course, when I reached the airport, I was also greeted with the natural elements of a thermometer that read -28-degrees, time to head home!

7:48

Fairbanks
Clear

-28°

Friday TODAY			-9	-28

Now	8PM	9PM	10PM	11PM
☾	☾	☾	☾	☾
-28°	-28°	-28°	-27°	-27°

Saturday	⛅	-20	-27
Sunday	⛅	-24	-30
Monday	⛅	-26	-33
Tuesday	☀	-29	-34
Wednesday	☀	-26	-27

What's Left?

There are many things in the USA that I would still like to see, but actually seeing all of them is probably unrealistic.

1. **Mt. Rushmore**, (Projected for 2020)

2. Famous National Parks such as Yellowstone

3. The Sky Walk on the East end of the Grand Canyon

4. I would also like to visit all 50 states, since Alaska was now number 47, it will just be the 3 M's left, Montana, **Maine & Mississippi**. No matter what it is that you like, there is always something to see in North America that offers beauty, history, and excitement.

3-South America

As I started my career in travel and thought about the many places I would like to see, the thought of taking a trip to South America was not something that I had in mind. Not that I had heard anything bad about traveling there, in fact it was really just the opposite. The owner of the agency where I first started to work had a regalia of stories about his trip to Peru, and how exciting his adventure to **Machu Picchu** was. Also, the lady who used to cut my hair had taken a trip to **Rio de Janeiro** and could not stop talking about how wonderful it was. Add to that the movie **Blame it on Rio** really had this young man's attention. Still, it wasn't a place I could see myself going, as I had much more interest in visiting Europe during this time. The one thing that I was starting to learn was that when an opportunity presents itself, make sure to take advantage, and that is exactly what happened when I found myself heading to South America for the first time.

Brazil & Peru
JUNE 1989

One of the best industry perks that were offered in the early days of my travel career were known as the Familiarization trips. These FAM trips, as they were called, were tours to destinations that were available to travel industry employees, with the sole intention of familiarizing employees, so that you could perform your job better. Depending upon the destination, would determine the type of trip that was arranged. Yes, these were travel trips, but they were also work in a sense. Often FAM trips included visiting hotels, checking out rooms, and sightseeing destinations. You were usually with a group of other travel personnel from different parts of the world, and the days could be very long. But these trips were also severely discounted, and in some cases, your company would pay the cost if they felt it was beneficial for their business.

I was working for a medium size corporate agency at this time called **Thomas Cook Travel.** They specialized in corporate account travel arrangements, but also offered leisure travel services as well. One day I stumbled on a flyer in the office that was offering a FAM trip to Rio de Janeiro for a 6-night tour. The tour included roundtrip airfare from Los Angeles, 6 nights hotel, a few meals and the sightseeing. The agent price for this tour was $329.00 for everything, and with that, I suddenly became very interested in South America. I had actually never taken advantage of a FAM trip before, but this seemed to be too good to pass up, so I went for it. Traveling to Brazil would also be the first time I needed to obtain an entry visit prior to traveling. Many people do not realize that when you travel internationally, you must have a visa to enter the country legally. In most cases, countries will issue you the visa when you arrive, and process you through immigration check points. The visa is the little stamp they give you in your passport, before obtaining your luggage. There are some countries however that require you to obtain an arrival visa prior to leaving your home country, and often there is a charge for the visa itself. Brazil was one of these countries, so I had to use a visa service, and obtain the stamp before I could travel there.

With my paperwork all in order, it was time to travel. Los Angeles airport usually provides non-stop flights to international destinations, so I was able to fly non-stop on **Varig** Airlines to Rio. Varig is the flag carrier of Brazil, so I was sure I would get the Brazilian experience from the moment we departed. This would be the longest non-stop flight I had experienced up to this point, which was 11:30 of flying time. The flight itself was pretty uneventful, and very smooth, which seems to be a by-product of 747's and larger planes in general.

The first day in Rio we were met by the tour host and went over some of the logistics about the trip, and what to watch for. Rio being a huge tourist destination had a reputation for petty crime, and pick pockets etc. Brazil is located in the Southern Hemisphere, so like Australia, traveling in June had me traveling technically in their Winter season. Though the climate in Rio is tropical all year round, it was considered off season. The balance of our first day was spent just checking in to the hotel and catching up on a well-deserved rest. Our hotel was the **Rio Othon Palace** which was not a particularly fancy hotel, but it was located right along **Copacabana Beach,** the most popular tourist area in Rio. There was an excellent view of the Harbor of Rio, which just happens to be one of the 7 natural wonders of the world.

Our first sightseeing adventure was to check out Sugarloaf Mountain in the harbor and included riding the cable cars between the two mountain points. This offered some amazing views of the harbor and the city of Rio de Janeiro as a back drop. The rest of the afternoon was free to check out the beach or the local shopping areas of Copacabana. It was during this time that a lady that was part of the tour had her wallet lifted right out of her purse, while she was still holding it. I felt bad for her because she wasn't negligent in any way, but it seemed the warnings we were given were sound. I immediately shifted my wallet from my back pocket into a front pocket for extra safety. That evening we had a reception at the hotel, and all seemed well, which should have been a flag to me, that something big was about to happen.

Through all the places I had traveled up to this point, domestic and international, I never had an issue with illness or food that did not agree with me. That was about to change when during my 3rd night there, I woke from a sound sleep, and rushed to the bathroom where I started to vomit continuously for the rest of the night. I could not think back to anything in particular that I had eaten or drank that would have caused this. I was very careful about drinking only bottled water and did not recall eating anything out of the ordinary. The one thing that I tried was the Churrascaria, which is a Brazilian concept where they slice different kinds of meat off of a skewer right at your table. Unless there was some type of bad meat, but it tasted very good at the time. The next morning, I was feeling a bit gingerly, and chose to just have bottled water, and some bland type foods like toast. Other people were eating eggs, bacon, and spicy sausage without any issues, but I did not want to take the chance, so I kept it basic.

Later that day, I wasn't feeling 100% myself, but I wasn't sick to my stomach, so I thought whatever I was doing it was working. We were doing some city sightseeing that day, so it was a lot of riding around, as opposed to walking. I made it through the day, and I was feeling better, or so I thought. A few people wanted to go out to a Brazilian club that night, and I made the mistake of having a beer. Again, around 2 or 3 am, I woke from sleeping, and again nauseated beyond belief. This time it did not go away in the morning, and by the time we were starting our day, I was still very sick. I could not sit out this day, because today was the day we were going up to Corcovado. This is the mountain top where the statue of Christ The Redeemer sits, and there was no way I was missing the most famous landmark of Rio. The ride up the mountain was by rail car, and not so bad, and while I was up there, I was holding my own. But on the return, we were on a bus, traveling on a bumpy road through the rain forest, and that made the nausea all that much worse.

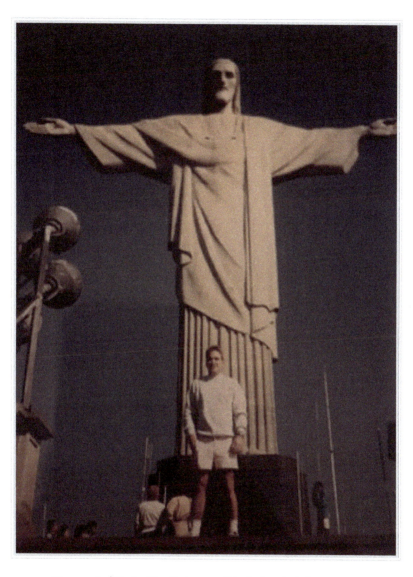

Corcovado (Christ The Redeemer, Rio de Janeiro), June 1989

I managed to make it back to the hotel in one piece but went straight to bed, still not really knowing what had affected me. I by passed on the last night farewell dinner, because I could not stand up straight, without getting sick.

I ended up sleeping from about 5pm until 8am the next morning and was fortunate that I did not have to wake up in the middle of the night again. We would have some free time to spend on our last day until we left for the airport for an overnight flight back. I spent some time just enjoying the beach and thinking about how much the Movie Blame It on Rio had exaggerated the topless beach concept. Yes, there were a few ladies topless, but certainly nothing like the movie portrayed it to be. About the only thing I was willing to blame on Rio was whatever made me sick.

The flight home was a bit interesting as we made a brief stop in the Country's Capitol **Sao Paulo** to pick up additional passengers. The vast size of Brazil can be deceiving, as we had a 1-day layover on the return flight in **Lima, Peru.** Even though Peru and Brazil share a border, the flight from Sao Paulo to Lima was still 5 hours long. Arrival in Peru was a bit sketchy as there were machine gun guards all over the airport. A sign of the political times in that country, as there was a huge military presence, and a constant threat of uprisings. After an extended layover, we were back on the plane, and headed for Los Angeles, a mere 7 hours away. As a top off to an already up and down trip, because I was arriving from a South American country, Customs and Immigration were especially stringent.

Though very grateful for the opportunity to travel to another continent, the concept of becoming ill while traveling probably jaded my overall impression of Brazil, and South America in general. I figured that would be my one and only trip down there, and as far as I was concerned, it you took a big pair of scissors, and cut across the Panama

Canal, and let South America float away, I would not be disappointed. One thing was for sure, I was never going back, or so I thought!

Rio de Janeiro
AUGUST 2005

After my 1st trip to South America, I had no real desire to go back there at all. The illness that I had experienced really clouded my thinking and made me become much more finicky when I traveled anywhere outside of the USA. That was OK because I didn't really think there was anything that would draw me back and given that I went the first time by pure happenstance, it didn't seem like a big deal to me.

In May of 2005, I was attending a Sales conference in Vancouver, BC where I met a fellow Amadeus employee. He was in charge of the South American Region for Sales, and was based in **Buenos Aires, Argentina.** He proceeded to tell me that no one in the region had any real experience with the software that they were now going to sell, which is what I was selling in North America. My leader at the time had suggested he talk with me and see if I could help. Help in this instance meant, could I come and teach a training class of the program, and how to position it in the market place.

My first thought was "South America, oh no", but then I realized that it was Argentina, and that was a chance for me to travel somewhere new. So, I agreed that I could help them in the future with this project. I was somewhat in the driver's seat because this was a business trip, and my company policy allowed me to travel Business Class based on the length of the trip. Also, since I was doing this for another region, their cost center would actually absorb the charges, therefore not affecting my budget. I agreed to do the training class in August, but as we got closer to the date, I was advised that the people needing the training were actually located in Rio de Janeiro. "Boom", the nightmare from 1989 flashed into my head. Would it really be as bad as it was 16 years earlier? I could not back out this late, but if I had to do this, I was going to make it work to my advantage.

When this trip came up, I was pretty much hording airline miles with **United,** and I was going to add a whole lot of them with this trip. A business Class ticket for 12,000 miles roundtrip, with bonuses for my Elite status with them, netted me almost 20,000 miles in my account. The flight down, and subsequent ride from the airport was straight forward. Once I got situated, I became very paranoid about what I would eat or drink, which consequently meant, I ate almost nothing. I could not get the vision of my last trip there out of my head, and it made me suspicious of everything I would eat. Even when I would eat at a nice restaurant, things made me very uncomfortable. I remember ordering scallops in butter sauce, and when they arrived, the scallops still had the muscle attached to them. Here in the US, the muscles are generally detached before serving, or at least the places I have eaten scallops at, serve them that way.

I was so pre-occupied with not getting sick, that I allowed that to prevent me from making the most of my trip. After the second day of training, I had the rest of the day, and the following day before my flight back to California was schedule to leave. I kept thinking another 36 hours here, and I am scared to eat pretty much anything, how would I make it? Then I had an idea about leaving earlier than planned. United had no flights before the one I was already confirmed on; however, Varig Airlines is an alliance partner with United, and they happened to have a flight that left for LAX at Midnight. I called them and was able to book a Business Class reservation on that flight, grabbed a taxi to the airport, and was on my way home a day early. This was so unlike me to leave an international destination early, but the fear of my previous trip there was just too strong for me to overcome.

The Harbor of Rio de Janeiro , August 2005

I know that my previous experience weighed much too heavy on me for this trip, and I also know that it is an unfair assessment of what travel to South America can be like. But the feeling of that time when I was ill, was just so vivid in my mind, that there was no way to block that out. In the end, it left me with a fairly jaded opinion of South America that probably still exists in me today.

What's Left?

After everything that I have written in this chapter, I don't think it would surprise you when I say it is the least appealing travel destination for me. Still, there are a couple of places that I would not mind seeing, even if they are not on the top of my list.

- Galapagos Islands, Ecuador: Beautiful islands located in tropical conditions near the equator. Much of Darwin's work on evolution was conducted in this area.
- Andes Mountains of Chile: Mostly for a skiing adventure that would take place during June – Aug
- Argentina: This is the Western Hemisphere's gateway to Antarctica. If I end up in Argentina, it will mean I am traveling to Antarctica, and most likely stepping on my 7th continent.
- Iguassu Falls: Located on the border of Argentina and Brazil. Twice the size of Niagara Falls.

The magnificence of the Great Wall of China (Beijing) , September 2017

A stunning view of the Potala Palace at night (Lhasa, Tibet), October 2017

A chance to experience the largest canine breed, the Tibetan Mastiff (Tibet),
October 2017

Standing on top of the world (Mt. Everest Base Camp, Elevation 17,056 feet),
October 2017

Mountain Biking in Nepal, October 2017

Family Vacation to the Grand Canyon ,June 2007

Cruising Mexico with my daughters, July 2016

The ultimate test of fear (First jump, Go Jump Oceanside, CA), Aug 2016

4-Europe

T he place that I have traveled the most extensively outside of North America is by far the continent of **Europe.** It was my first International trip besides the Caribbean, and the continent that I have visited the greatest number of times. Through a combination of pleasure and business travel, I have pretty much traveled all across what we would consider to be Western Europe. Using just about every mode of transportation available, I created a series of adventures that some people might not find enjoyable, but to me they are forever ingrained in my memories. I managed to do most of my travel to Europe during my single, and newly married times of my life, which was probably best, because I am not sure I would undertake many of these adventures at this stage of life. So, sit back and let me take you through the most diverse continent in the world.

Great Britain
October 1986

I had always known that I wanted to travel to Europe at some point in my life but wasn't really sure when I would have the chance. I had been working in the travel Industry about a year, but did not make a lot of money, and even with the discounts, Europe could still be a pretty expensive trip. The one thing there was never a shortage of in the travel industry were vendor events that you could attend. One morning I attended a breakfast event offered by Pan American. The event was to celebrate their recent announcement for starting airline service to **Shannon Ireland.** So, you have a free breakfast, and you get to sit and listen to a speech telling you why you should sell seats on Pan Am's new service. This was all fine and dandy, but make no mistake, the real reason for attending these airline events was for the door prizes. Usually they will have a few drawings for free tickets etc. Whatever power you like to cite (luck, divine intervention), it was with me that day because my business card was pulled, and I won a ticket to Great Britain or Ireland, space available.

I was really excited about winning the ticket, but Ireland has never been on my list of "must see's", and in fact to this day, it remains one of the few countries of Western Europe I have not been to. I was much more interested in visiting Great Britain, particularly England, but also put together plans to visit Scotland as well. Now if I only had that Scottish lady from Princeton's address or phone number, I could have contacted her. When I met her 10 months earlier, I would have never thought I would be traveling to the place she was from.

Because I was traveling on a Space Available ticket, I knew I had to choose dates that were off season, otherwise I risked not getting on the flights. I decided to go in October, which hopefully meant the Summer tourist season was done, but the weather would still be mild enough to enjoy. One of the drawbacks to traveling on industry free tickets is the airline policy of being dressed in business attire. That meant for men a suit and tie while flying, and they would deny boarding if you were not properly dressed. So, there I was in my suit, holding my ticket with the words **"Upgrade to "J" at gate"**. That was an industry code for upgrade this passenger to Business Class upon check in. For a 6-hour ride across the Atlantic in Business Class, I would gladly wear a suit and tie.

The flight over was extremely smooth, and before I knew it, it was the next morning and I was on the ground at **Heathrow Airport.** Coming from the USA, customs and immigration was a breeze, and I soon found myself on the **Piccadilly Underground** line taking the train into Victoria Station. Many people will tell you that you should experience the double decker buses or cabs in London, but I found the London Underground to be much more efficient, considerably less expensive, and much more fun. My first day consisted of resting, and general sightseeing walking around. I was able to walk from my hotel to see **Buckingham Palace, Westminster Abbey, Tower Bridge, Big Ben, and the Houses of Parliament.** It was a little confusing when crossing the street, you need to remember to look right first, because they drive on the left side of the road. I have always hated when people say, "they drive on the wrong side of the road". It's the left side, not the wrong side, but I digress.

After spending a couple of days checking out London proper, it was time to branch out into the country. You need two things when traveling to Great Britain, an London Underground Pass, and a Brit-Rail Pass. Those two things will pretty much get you to wherever you want to go. I decided to take the train out to **Oxford University,** yes that Oxford. About an hour's train ride outside of London is the town of Oxford, with its Harvard type university. Very prestigious, and located in a quaint little town, Oxford is small streets, small shops, and lot of pubs. 31 years later, I still have the souvenir coffee cup listing all 26 colleges in my cupboard. It made for a great day, and in the afternoon, I headed back to London. The next day it was recommended to me to visit the city of **Bath,** which was about 2 hours West of London. The only thing I knew about Bath was it is the home to the band **Tears for Fears,** but on the train I went. What I found was a city with amazing Roman architecture and many natural hot springs. The Roman Empire did occupy that area for a period of time during their rule of power. My final day in England had me heading toward the coastal city of **Brighton** in search of some authentic English fish and chips. I found a shack on a small pier, and sure enough I was indulging in battered fish, chips, and pint of beer, served at room temperature.

I could have spent many more days taking day trips from London to some other interesting places like **Stonehenge, Stratford Upon Avon, and Liverpool,** but I just didn't have that much time, and I wanted to see something a bit different. With that in mind, I decided to take a couple of days and travel North to **Scotland.** Having my trusty Brit-Rail Pass, I boarded the train from **Kings Cross station** heading North to **Edinburgh** (pronounced Edinboro), the Capitol of Scotland. It was about a 6-hour train ride with brief stops along the way, in the spacious first-class compartment. Other than the train running about an hour behind schedule, the ride itself was pretty uneventful. I met a little boy named Tony who was riding alone, without an adult. He could not have been more than 10 years old but seemed quite comfortable being on his own. It was funny when he said to me with his English accent that he needed to go to the **toilet.** There is just something about that word "toilet" in American culture, that makes it sound so dirty and gross, yet it is the common expression around the world, much more so than Bathroom. Something interesting that I noticed as we pulled into the Northern English city of New Castle. The accent of people started to get much heavier, and in some cases, made it darn near impossible to understand English. This is what I would find in Scotland, as the Scottish brogue can be very difficult to understand.

Pulling into Edinburgh the weather was not cooperating as it was raining and would pretty much remain that way for the 3 days I was there. Not daunted by the weather, I was able to check out Princess Street, and then walk up for a tour of Edinburgh Castle. Along the way, I saw a traditional street player in full kilt, playing the bagpipes. I was hoping he might break out a little ACDC "it's a Long Way to the Top", but no such luck. Doing a little bit of shopping, I came across some hard to find VHS concert videos that I wanted to buy. My international naiveté just assumed that VHS was VHS, never knowing that the American system runs on a platform called **NTSC,** while everyone else runs on what is called the **PAL** system. When I returned home, the tapes would not play in the VCR. I was very fortunate that the company where I bought the tapes, did refund my money, after I returned them via mail.

The weather prevented me from exploring the Loch Lomond area, where the myth of the Loch Ness Monster originates from. So, after spending 2 days in Edinburgh, I actually decided to return to London, and take care of a billing snafu by the hotel I had stayed at. It seems they over-charged me about 65 Pounds, so I returned and let them correct that error. Soon it was time to head back out to Heathrow airport for my flight back to JFK. Traveling on a Space Available ticket is such an uncertainty, that you have to have a backup plan, so heading home a day earlier wasn't a bad idea. As it turned out, I got the last Business Class seat on the flight and made it safely back into America. One surreal note to this moment is my return flight was Pan Am flight 103. If that doesn't ring any bells, 2 years later in 1988, **Pan Am flight 103** would be blown up by a bomb over Scotland, in what is known as the **Lockerbie Bombing.** It would not be the last time that traveling to Europe had me close to historical world events.

Upon my return home, I presented my mom with some souvenirs of the English Monarchy. She has always been into that Royal Family stuff, and there was certainly no shortage of it available to buy. It was a way for me to make her a part of something that I knew she would never see for herself.

My overall trip was amazing, and I got to experience for the first time, a true sense of international travel. I found the English people to be very friendly, and very helpful when needing directions or advice. My experience with English food was not all that spectacular, as I found many things tasted somewhat bland, but still well worth trying Bangers, and Mushy Peas. I love peas but there is something so unappetizing about them being mushed into a ball. Finally, I found that I was quite at home running around another country on my own, doing what I wanted to do, and not being afraid to venture or try something new. A spirit that would launch my desire to travel around the world as much as I could.

Germany, France & Austria
September 1989

The success of my first trip to Europe really paved the way for me wanting to travel there much more extensively. I started to become much more interested in my German ancestry and started taking German language classes through the local Community college. It was time for me to experience more of Europe, but I wasn't really sure what I wanted to do, besides visiting Germany. Then an unexpected event took place that would delay my next trip to Europe. I moved from Pittsburgh to California in 1987, and it would take me a year or so before I got on solid ground to be able to take some more exotic trips. I had heard stories of people when they are young, often backpack through Europe as an inexpensive way to travel. I didn't think that was me, but as this story goes, it was darn near the backpack experience I had heard about in terms of doing whatever it took to get from point A to point B. From early on in my traveling days, I always liked to try things that were just a bit different than the typical traveler. Yes, I like to see the tourist places, but I also like to try things where not everyone goes, or even has the desire to go. That is when I came up with the idea of traveling to one of the most famous festivals in all of the world, the **Oktoberfest in Munich Germany.** Yes, there are versions of the festival around the world, and especially in America, but this was the real one. This is the original event to celebrate King Ludwig's marriage in 1810, when all the people were invited to the party. Since this event is held during a certain time annually, I had to make sure I could schedule the time off.

The first thing I discovered which most people still don't understand is that Oktoberfest for the most part takes place in September. It starts the 3rd Saturday in September and ends the first Sunday of October. Now knowing the dates, I had to plan my vacation accordingly, and check flight availability. Then the magic of discounts kicked in as Delta airlines was offering a special on their new service to Germany, which was $175.00 roundtrip, confirmed space. Man, I miss those special days, as they no longer exist. I didn't really have much of a plan other than I was going to be in Munich on certain dates for the fair, but the rest was play it by ear. So off I went as my journey started for the German city of **Cologne.** I wanted to see the Gothic Cathedral, the only structure not damaged during World War II. Once I arrived, I saw the Cathedral, checked out some sites in the rest of the city, and said to myself, "this is it"?

There wasn't a lot there that captured my interest, so it was time for a little impromptu action. I pulled out my discounted **Eurail Pass,** as it was called back then, and looked at some time tables, and decided to adlib a bit. I noticed a train the next morning that could have me in **Paris France** by early afternoon, but I had to be aware that I needed to be in **Salzburg Austria** in 2 days for a hotel I had confirmed there. I had never intended to go to Paris on this trip, but I had also never been there, and this was a great way for me to hit another new place. So, I got up early the next morning, got to the train station, and I away I went. Crossing through Belgium, and into Northern France before arriving into Paris 5 hours later. I was traveling with a big bulky suitcase, so I found a locker at the station "**Gare Nord**", (the station name will be important soon), and placed it in the locker. Now I was free to roam about the City of Light, not having a clue where to go. I had about 10 hours to kill, and I was going to make the most of it. I picked up a free city map, changed some **German Marks** into **French Francs,** and was on my way walking the streets of Paris.

Hunger kicked in, and I happen to find a McDonalds right there, and yes, I knew the Quarter Pounder was called a **Royale** in France, long before **Pulp Fiction** was ever a movie. As I continued walking I made it to **Notre Dame, The Paris Opera, The Champs Elysees, the Arc De Triomphe, the Plaste de Concorde, and finally to the Eiffel Tower.** As nightfall hit, I was able to take a **Seine River** cruise with the city all lit up, a trip I would take several times again on future trips there. I don't really know how many miles I walked that day, but being an unseasonably warm day, made it that much more enjoyable. Now it was time to head back to the station to retrieve my luggage and take a train down to Salzburg Austria.

I had spotted a train that was scheduled to leave Paris around 11p, so I got to the station with plenty of time left, got my luggage from the locker, and soon realized my 1st major screw up. I arrived in Paris at Gare Nord, which is the same as saying the Paris North station. The train for Austria left from Gare Est, the Paris East station. I was soon running through the streets of Paris, with a big suit case trying to get to the other train station, all the while looking at the map, and trying to read French street names. Somehow, I made it with a few minutes to spare, but then I discovered my 2nd major screw up. Generally, you do not need to make train reservations on most trains, however there are some peak exceptions. The train I was taking to Salzburg was going via Strasbourg France, which happens to be a huge military base area. This was a Sunday night, and all the soldiers were returning to work on Monday in Strasbourg. Needless to say, there were no seats available on the train, and people just kept packing in. I boarded as well and slept on top of my suitcase for the first 5 hours of the journey in the aisle of the car. All night long people getting off and on the train were stepping over us like we were cattle. I certainly did not get any sleep that night, but I was fortunate after the train reached Strasbourg, seats opened up, and I was able to rest a bit better. This was probably my first roughing it experience, and while not pleasant, I look back now and laugh at what I had to do to stay on schedule.

A 12-hour train ride later had me arriving in **Salzburg Austria** very tired, but no worse for wear. Though I did not know this yet, the Alps Region would become one of my favorite places to visit. I chose not to go to Vienna, because quite frankly, I had my fill of the big cities, and wanted to experience the small Tyrolean feel of the country. Salzburg is probably most famous for the setting that would become the story of **The Sound of Music**, but it is also a gateway to some of the best skiing in Europe, especially in Austria. The architecture was pretty much as advertised, small village shops with coffee and pastries everywhere. Narrow streets made of cobblestone giving way to parks and fountains. After a good rest, I decided to take a full tour, and not coincidently, it was called "the Sound of Music" tour. They basically took you on a tour of all the locations that were used in the film, and the bus had piped in soundtrack music. This was total tourist, and I loved it because I would have never thought to check out all these places myself. Of special note, the church that is shown in the film is nowhere near as big as it is made to look in the film. In fact, The Von Trapp's were not even married in that church. All in all, a great couple of days in a beautiful mountain town, but now it was time for why I made this trip in the first place.

Having some training in German language made things much easier to get around in Austria as I did not need to ask for directions the way I did in Paris, and now, I was going to become even more immersed in German, as I settled in for a brief 2-hour train ride from Salzburg to the Capitol of Bavaria, **Munich.** Munich was the second largest city in what was known then as **West Germany,** but being in the mountains, it also had the alpine feel. This is the home of the Oktoberfest, and beer tents, and just a whole lot of fun. I somehow lucked out, as the tiny little place I was staying (only 10 rooms), was only ½ mile from the fairgrounds where the Oktoberfest is held. I envisioned this event to be a lot of beer tents, food, and music, but much to my surprise, it had an amusement park element to as well. Carnival rides, and games were as much a part of it as well and caught me off guard a bit. The event was grand in size, but now it was time to partake in some of it. Of course, there were liter steins of beer being carried by Frau's 8 at a time, but then I discovered a little piece of Heaven on Earth. 18 inch giant hot pretzels made fresh with just a touch of salt. These things were a meal in themselves, and you could just sit back, drink a beer, and listen to the Umpapa bands play **Ein Prosit.** Such an amazing night, and sadly, my lousy camera took terrible night pictures.

I allowed for a full day in Munich to see the sights, but this time on my own. The **U Bahn & S Bahn,** which is the subway system, are very efficient, and I wanted to experience that instead of tour buses. I checked out the city center of the **Marienplatz,** with its famous Glockenspiel, and dancing figures where everyone stops to watch. Visiting the original **Hofbrauhaus** for a beer and lunch, and just enjoying the German culture. Germany of course has an infamous past, but I think the German people have paid a steep price for that, and now they just focus on having fun, and enjoying life. No matter where you go, there always seems to be sausages on the grill, and beer in the stein.

I really didn't have any plans on my last night in Munich, and that's when I usually get creative. Out came the train schedule, and I noticed that the Winter Olympic City of **Innsbruck Austria** was just a 2-hour ride away. This was a no brainer, and a new city to see, so down to the train station I went. On the train, I met a group of Americans who saw me sitting by myself and asked if I would like to join their group. They were all flight attendants with American Airlines and were given several days off because

Hurricane Hugo had struck their home base of **San Juan** hard. So, they decided to come to Munich for a few days.

Innsbruck was just what you would expect from a tiny mountain village. It is hard to imagine that this city has hosted several Winter Olympic events, with the most recent being in 1976. We had dinner at a little inn and walked around the city looking at handmade crafts. It was a quick trip to be sure, but it also was a perfect example of how compact Europe is, and how you can quickly get from one culture to another in a short period of time.

My departure from Munich back to America was on October 9th, 1989, and while that day is not really anything of significance, I missed one of the most monumental events of the 20th century in Germany by one month. November 9th, 1989, the Berlin Wall came down thus starting what would lead to not only the re-unification of Germany, but of leading to independence in all of Eastern Europe from Communist governments.

A memorable trip for me not only from a personal perspective, but from a historical perspective as well. I knew I wanted to visit the area again, but I had no idea I would be back in less than a year from this time.

European Honeymoon
AUGUST 1990

Having taken so many amazing trips up to this point, you might get the notion that everything was perfect. Yes, I had some amazing adventures to both Europe and the Caribbean, and even tossed in a few domestic trips that defined my life. Still, each of these trips to this point had one thing in common, I took them alone. All of these grand trips, and no one to share the experiences with. That would change with my engagement, and subsequent marriage in July 1990.

I know many couples opt for the exotic paradise beach type of honeymoon, but I really wanted something special for this trip, and Mary had never traveled internationally before. We didn't really talk much about where to go prior, and since she had never been to Europe before, I thought why not do it all, or at least as much as we could. The stars of my career continued to align as 9 months prior to our wedding, I accepted my first leadership position with a tour company, that specialized in trips to Europe and Australia for ages 18 – 35 years old. One of the company tours became the basis for what would be a 3-week tour in Europe, that would touch 8 different countries.

I was finally going to get to share the experiences I had with someone very special to me and take advantage of the opportunities working in this industry gave me, as we basically traveled this whole trip for about 10% of what the public would pay. Since the tour had scheduled departure dates, we actually waited about 10 days after our wedding day, before we left for our honeymoon. That really turned into a good thing because when you travel abroad, and for that length of time, so many details needed to be taken care of.

Departure day arrived, and so did we at LAX for our nonstop flight to London on Pan Am (I miss them). I had not said anything to Mary at this point, but our airline tickets had the endorsement "Upgrade to "P" at gate". This wasn't Business Class, but First Class, as the letter "P" was code for Premium. This was going to be a surprise assuming we cleared the flight, and a surprise it was for the 10-hour jaunt across the Atlantic.

One travel tip that is always recommended is that when traveling to different time zones, always try to stay on a normal schedule. We were now 9 hours ahead of California time, and it was mid-morning in London, so ideally, we should have just kept going, but fatigue set in, and we rested when we got to the hotel. Later that day we started our sightseeing in London, as I took Mary to all the tourist spots, Buckingham Palace, Westminster Abbey, Big Ben, Parliament etc. After a full day, we found our biological clocks off balance and were wide awake at 3am, but hey, we were on our honeymoon, so we found some things to do.

We spent the next couple of days in London sightseeing before our tour of the Continent started. A trip to **Windsor Castle**, and some other cities was in order, and it gave us a good feel for what we could expect from the tour. Europe was experiencing a Summer heatwave that year, and many places did not have air conditioning, so it was a bit uncomfortable. I remember seeing some businesses advertising that they had air conditioning as a way to entice customers. After 3 days in London, it was time for us to start the escorted tour, and for the next 16 days, we would visit 7 more countries, make 40 new friends, and meet one Wally!

At the start of our escorted tour, we would board the bus in London, and drive to the coast for a ferry ride across the English Channel from **Dover to Calais**, France. It was only about a 1-hour ferry crossing, then another bus would pick us up, and continue the land journey to Paris, where we would spend 2 nights. Since I had just been to Paris, less than a year earlier, many things were still fresh in my mind. I took Mary to most of the sites, The Eiffel Tower, Notre Dame, Arc De Triomphe, and we took that night time Seine River cruise. Our second day was spent visiting a few Paris sites, and then the bulk of the day was at the **Louvre Museum**. This was my first time at the famous Paris museum, and I was anxious to see some of its landmark exhibits. The **Venus De Milo** I thought was a bit disappointing, and the biggest attraction, **Da Vinci's Mona Lisa** painting, was somewhat anti-climactic.

While our tour started a couple of days earlier, the structure of the first couple of days didn't lend itself to actually meeting people on the tour. This would take place on day 3, as a culture of Australians, English, Canadians, South Africans, and Americans made up the group. We were scheduled to leave Paris, but suddenly we were delayed because two people on the tour were missing, and we could not leave without them. About 20 minutes later, one of the missing guys appeared, and would become a very important part of my life. His name was Lindsay, and he was from Australia, traveling on his own throughout Europe. As we rolled South toward the **French Riviera,** the tour guide thought it would be nice for everyone to stand up and introduce themselves. One by one we each got up and gave our name, and where we were from. Lindsay feeling a bit embarrassed, stood up and announced everyone already knew who he was because of his lateness. Then an English gentleman stood up, and introduced himself as David, however, it seems many of the Australians had given him the name of Wally. It turns out that in Australia, the name Wally would be symbolic to the name **Eugene** in America. Often depicting an impression of a nerdy looking person. Though it was a bit of a put down, David was actually a good sport, and even sort of played along and was referred to as Wally for the whole trip.

We spent a night in the city of **Cannes** and took a trip to the **Principality of Monaco** to the casinos of **Monte Carlo.** The 3 casinos were very high end, and for the very wealthy, but it was well worth having a look. Before heading back to the bus, a group of us stopped off at small little bar for a beer. Soon the rest of the tour showed up there, and before we knew it, the juke box started playing the classic **Chicken Dance song,** and we all stood up and started dancing to it. Finally, a few of us went for a midnight walk along the beach enjoying the Mediterranean Sea, and the lights of Cannes.

The next 5 days would be spent in Italy, and would see us stop in **Florence, Rome, and Venice.** The long bus rides were conducive to sleeping and getting to know fellow tour members. Mary and I started talking to Lindsay more, and we began hanging out at the stops, and night time fun. Stops in Florence allowed us to see the **Leaning Tower of Pisa,** The Piazza St. Croce, and a visit to the Academia to see **Michelangelo's statue of David,** one of the most famous sculptures of all time. A couple of nights in Rome had us seeing **The Colosseum, Circus Maximus, The Spanish Steps, The Trevi Fountain, and of course Vatican City.** It was very warm in Rome at this time, so to ease the heat, we would visit a small bar across the street from our hotel, which also moonlighted as a gas station. A gas station that had a small outdoor bar was officially dubbed **Gas & Sip,** by the crew. While we were all having fun, we were reminded that we were traveling in another country, and caution must always be exercised. A couple of ladies, including my wife, wanted to return to the hotel, and their short walk across the street was greeted with some local men following them into the parking lot, and harassing them. Fortunately, nothing came of it, but I know she was startled by what happened.

Early morning visit to the Colosseum (Rome, Italy), August 1990

Early morning departures were becoming more difficult as they were preceded by late night bar stops with
Lindsay. It was right around here I started thinking how
I wished I had someone to run around Europe earlier in
my life like this. Now I was married, and felt I had to be-
have like a married man, and not some single guy carousing
around Europe. Lindsay also had found a female playmate
on the tour, who began following him like a lost puppy.

As we headed north to Venice, some more rest was in
order, along with some aspirin. Venice would be a new location for me, and I was quite excited to see what the Water City was like. The city was very much like any pictures
you had seen, but photographs do not tell you the full story
when it comes to scents. The water in Venice is very pol-
luted, and had a really foul smell to it, still it was Venice,
and gondolas and motor boats abound. Some of the most

beautiful hand-blown glass objects were located here, but extremely overpriced, which at the time much of Italy had been. The dollar was not strong in Europe, and it made budgeting a constant challenge.

A couple of days in **Vienna Austria** made for a bit slower pace and reaffirmed my decision from the previous year to by-pass Vienna on my trip to Germany. Vienna is a nice city, but just that, a large city, but nothing that I had a huge desire to see. The lull would only last a couple of days, because the next stop was going to be Munich. I was glad to be back, and refreshing my German language made it that much more special. A night out at the Hofbrauhaus was a must, and after a few of the Litre's, I summoned up the courage to guest conduct the umpapa band. There I was on the stage with my conductor baton, but which song to lead? There could be only one choice as I looked at the band and said "**Vogeltanz**", translation, Chicken Dance. As we started, the crew stood up, and away they danced to what became the official party song of our tour. My second trip to Munich was just as fun as my first, only this time I was sharing the experience with someone else.

A chance to lead the Umpapa Band (Munich, Germany), August 1990

All of this running around from country to country was the perfect scenario for catching a Summer cold and catch it I did. Of course, the late-night partying, and the early morning departures did not help the situation, neither did the fact that we were heading through the Alps, at higher altitudes, and cooler weather. Our destination for the day was **Lucerne Switzerland,** and we arrived there mid-day, after a drive through the tiny country of **Lichtenstein.** The mountains of Switzerland were probably the most scenic part of the tour, as green hills gave way to snowcapped peaks. Dinner tonight would be at a famous restaurant located at the top of **Mt. Pilatus** in Lucerne. The next morning, a stroll along the lake for shopping led to bucket list purchase, and one of the few items we actually bought. I have always loved authentic Cuckoo Clocks, and this was my chance to get one. The time pieces are made in Switzerland, while the housing cases are usually made in the Black Forest region of Germany. We picked one out and arranged to have it shipped back to California before our departure from Lucerne.

Having lunch in Lake Lucerne, Switzerland, August 1990

Now it was time to travel back through the Rhine Region of Germany that would take us to a little town called **St. Goar**, along the Rhine River. It was a small little village, and our hotel here was actually a small guest house where our group were the only guests. This place became jokingly known as the Hotel Eingang. The word **Eingang** is actually German for entrance, and the sign above the door said Hotel Eingang, meaning the hotel entrance. The group had a lot of fun referring to it as the Hotel Eingang. Since it was a small guest house, the bar consisted of a few chairs, and a few draft handles. As most people went to bed, including the bartender, Lindsay and I slowly finished our beers, then decided to pour our own. This was probably the most inebriated I was for the whole trip, as we talked until probably 3am. When we both decided we were done, Lindsay thought he would visit one of his two friends he was romancing on the trip. On his way to one room, the 3rd member of the triangle was looking for him as well and were headed for a showdown. I managed to duck into my room, where my wife had been sound asleep for several hours. Somehow, he managed to get out of that situation, though he didn't remember how either.

Clearly feeling the effects of my own hangover, I settled in for a long bus ride to our final stop on this tour, which would be the city of Amsterdam, in the Netherlands. I had always heard about how free and loose this city was, and I was going to get to see for myself. The culture in Amsterdam is much more open than we are used to in America. Hash bars, and small amounts of controlled substances are legal, and available. Prostitution in the Red-Light district is legal and is part of the allure of the city. After dinner, a group of people talked about attending a sex show in the Red-Light district. Mary said to me that she wanted to check it out, and I was surprised because she has generally been a pretty conservative individual, but hey, this was a chance to experience something very different. My only stipulation was that these shows are pretty expensive to get into, around $50.00 per person, so if we go, we stay and not get offended after 5 minutes.

So off we went to check out the seedy side of Amsterdam, and the show did not disappoint. It was a full nude show with costumed characters including Batman, and pole dancers. The highlight was when Batman danced into the audience and approached our group of ladies and presented them with the Full Monty about 2 feet away. The look of shock was present, and some had the look of desire, but it was all in good fun, and certainly an experience that we didn't forget.

Our final day was a 6-hour ferry ride from Amsterdam back to England for the end of the tour. It was a chance to prepare to say good byes to a group of people, who in some cases became almost like family. Many of the group were staying in London, or traveling on, but for Mary and me the next day was the long flight home. It was hard for me to say good bye to Lindsay, after all the things we experienced and shared on the tour, and I immediately thought about how to arrange a visit in the future. As it turned out, there would be several visits and many communications in the years that would follow, leading to an amazing long-lasting friendship.

We had one last thing to do, and that was our flight home, but hey, it was first class, and it also happened to be my birthday, so what could go wrong? How about NO SEATS available on our flight to Los Angeles? Every seat, not just First Class, but every seat was taken, and we were not getting on. I quickly shifted into Travel Agent mode for emergencies and tried to figure out what to do. We tried the flight through JFK, sold out, the flight through San Francisco, sold out. Are we really going to get stuck in London for another night, or is there one more rabbit I can pull out of the hat? There was one rabbit left, as I ran down the concourse and found 2 seats available on the flight through Washington DC, that would get us home much later, but we would still make it. The stress of barely making the flight was made much more tolerable as we sat down in First Class, and before we stowed our carry-on luggage, the flight attendant had 2 champagnes waiting for us. A few more after takeoff, and it was time for an extended nap, and time to take stock of the most amazing honeymoon trip that any couple could be so fortunate to take. I finally got to share an experience with the most important person in my life.

Scandinavia
May 1991

As I mentioned in a previous chapter, there are times when a traveling opportunity comes along that you just never thought would be a place that you would want to go. For me that opportunity came in the Spring of 1991, but actually started in December of 1990. I had written previously how my first trip to Europe came about by me winning a prize drawing at a local airline event. I had been to many events where prize drawings occurred, but never really won anything since that first time. That was about to change as I attended the American Express Christmas party for the Los Angeles area in December 1990. I had only been with the company for a few months and didn't really know many people outside of my own office, but decided to attend the party, and it was a good thing I did. My business card was drawn for 2 American Airlines tickets to **Stock-**

holm Sweden, that I had one year to use.

The Scandinavia area is a place that never really drew my interest as a place I had to go, but when the transportation is on the house, your attitudes tend to change. No matter where you visit there is always something to see or do, and I'm sure this was no different, and I just needed to do a bit of research to make this a memorable trip. I immediately knew since we were going to Sweden, we needed to include some other Scandinavian countries as part of this trip too. Mary pretty much let me plan this adventure out on my own and would be willing to go along with what I thought would be best for us to see. So, in addition to Sweden, we also decided to include **Denmark and Norway** as part of this Scandinavian holiday.

We chose to go in May of that year because of the Summer rush that Europe experiences, but also to make sure the weather would be mild enough to enjoy the trip. Our trip would start In Stockholm, Sweden for the first 3 days. When we arrived, we discovered the International Airport is a significant distance from the city, so after a long flight we still had a long ground ride to our hotel. Stockholm was a very clean city that is surround by a lot of water in the form of rivers. Touring around the city, and seeing a few museums was the bulk of our first day, and while the experience was nice, I soon realized that this was a place devoid of many of the famous landmarks and architectures of other European cities. One aspect that was similar to many other European Capitols was the everyday cost of items. Sweden as a whole was a very expensive country, and the strength of the dollar at this time, made matters even worse.

Our second day was setup to visit the **Drottningholm Palace** which is the home of the Swedish Royal Family. The palace actually dates back to the 16th century, and while it is still used as a residence, it is also open as a tourist attraction. While this visit made for a nice day, I had started to become palace weary having seen so many palaces throughout Europe, as such, they all started to look the same.

One thing that I thought would be a bit different from our other trips to Europe was to book an overnight train from Stockholm to our next destination, which was **Copenhagen** Denmark.

Taking the overnight train, I made sure to book us a sleeping compartment. The memory of my Paris fiasco a couple of years earlier, and not wanting Mary to experience that type of a trip, I made sure to reserve a private sleeping compartment for us. Though the compartment was private, I think she would have preferred a regular hotel for the night's accommodation.

An early morning arrival in Copenhagen was met with quite cool temperatures, and a wait for our hotel room to be available. Of all the cities in Scandinavia, Copenhagen seemed to be the one that I had heard the most positive things about. One of the places that I was told you had to visit was **Tivoli Gardens** in downtown Copenhagen. I was a bit surprised when we got there because I had a pre-conceived notion of what Tivoli was. Just like the Oktoberfest was surprising to me, so was Tivoli in that yes it had beautiful gardens, but it is an amusement park, one of the oldest ones in the world. Being there in May, it was only open on a limited basis, and the crowds at this time were very small.

From a historical reference, they were a couple of things in Denmark that I had fixated myself on seeing, and of course, Mary was up for it as well. We took a tour out to **Kronborg Castle** in the town of Helsingor, and yes it was another castle of the many we had seen, but this one was the setting for **Shakespeare's Hamlet.** That's right, Hamlet's Elsinore Castle is actually Kronborg Castle. On our return from the castle, we had a stop at my most identified tourist spot in Denmark. Ever since my early days as a leisure travel agent, I would always see in brochures that featured Denmark, the famous Little Mermaid of the Seas statue. The Mermaid sits on a stone in the sea as a guiding point for all the ships, or so the pictures would have you believe. The reality is, the Mermaid is a relatively small statue that sits on a rock along the shoreline. It's not a big statue in the middle of the water, and on the whole was a big disappointment for me.

After 2 days of sightseeing in Copenhagen, we had one more full day available before we took another overnight train to Norway, and no real agenda. When in doubt, pull out the Eurail schedule and see what new, unplanned adventure we could do. 2 choices came up, the first would be to take a ride out to one of the outland islands of Denmark to visit Legoland. That's right, the original **Legoland** amusement park is built right next to the Lego factory. The second option was to take a train ride down to Northern Germany and visit the largest city in West Germany, **Hamburg.** Re-unification was still a few months away, so there was still an East & West Germany at this time.

Given the weather and the ease of travel, we opted to

spend the day in Hamburg. We really didn't have a whole lot of time to explore but managed to grab a quick city tour around the waterways. At the end of the tour it was time to grab another train, only this time our final destination was **Oslo Norway.** This would be another long train ride, and this time there was no private room, but rather sleeping couchette's, 4 to a room. It was not surprising that this would be the last train travel that my wife and I would take on our travels, as she was not impressed with this much adventure.

Traveling to Norway was probably my least amount of familiarity of all the places I had visited. I was aware that the famous Fjords of coastal Norway were sights to behold, but we had not planned on going that far inland based on the time we had. The city of Oslo was a very clean city, and designed very much like Stockholm, without the hustle and bustle. We stopped to visit **Vigeland Sculpture Park**, which is an entire outdoor park dedicated to large sculptures. Historical, heroic, and even a section of erotic sculptures was available for your viewing pleasure. The most famous resident of this park is a statue of an angry young naked boy named **Sinnataggen**. His scowling face, and raised arms make him the park's premier tourist attraction.

One thing that I vividly remember about Oslo is that because of its Northern Geographic location, daylight in mid-May lasted until midnight. This was probably the furthest North I had been on any of my previous trips, including Scotland. Our final day in Norway had us visiting a small-town north of Oslo, which was the village of Lillehammer. Back in 1991, Lillehammer did not mean anything to me, however, in just 3 years, it would be the site of the 1994 Winter Olympics. The same Olympics where we experienced the **Tanya Harding/Nancy Kerrigan** saga. When we were there however, much of the development was just getting underway, so it was still a relatively small town.

As we were returning from out 9-day trip to Scandinavia, I realized that it was another place that I had been to, but probably by far, my least memorable trip. That is not a bad thing, or a good thing, but just a trip that did not leave me with any lasting memories that I would one day be anxious to share the stories of. Still, when you have the opportunity to visit any new country, regardless of what your pre-conceived notions maybe, I absolutely recommend to do it!

London
June 2002

It had been 11 years between trips to Europe, and so much change had occurred in my life between that time. I now had 3 daughters and was 12 years into my marriage. There is no doubt when you decide to start a family, that takes precedent, and the exotic long vacations seem to disappear. So it was with me, but something else changed as well, my career took off on the technology side. I now had completed a master's degree, and that academic hard work was starting to pay off. I had gone to work for an Oracle subsidiary as a Project Manager, and this led to a significant increase in business travel for me. I was assigned a project where the company had offices in the suburbs of Boston, and also in London. When we started a project, we would often gather all parties involved for a kickoff meeting designed to introduce the team and outline the scope. The customer wanted to have the kickoff meeting in London, which was a first for me to travel abroad for business.

I decided to take an extra couple of days the week end prior and head over to London early, so I could get my tourist things in. Business travel is often for just a couple of days, and since it had been so long between trips, I wanted to make the most of the time that I had. When I left, I was woefully unprepared with the little things, like voltage converters, currency etc. Having not traveled abroad in so long, I simply forgot those little things that can turn a trip from success into disaster. The last time I was in Europe, there were no cell phones, laptops, and all the other electronics that are common place now. I actually felt a bit pensive and scared about being in another country. I guess it took me about 2 hours to get my bearings, and comfortable again, and then started my scavenger hunt for the things I had forgotten to bring.

Besides the usual tourist sights, I had seen before, I went walking to re-kindle some of those old memories, and to prepare myself for what was to come that evening. There were some things that had changed in London, and some things that stayed the same, like the food, bland as ever. Between walking and a lack of desire for eating, a trip to London always made for a little weight loss. None of that mattered because I was finally going to get to do something

I have wanted to do ever since my first trip to London 16 years earlier.

For as long as I can remember I have had a morbid curiosity for the events of **Jack the Ripper.** The myth, the theories, the brutality of it and the unresolved identity have long fascinated me. Now I was going to address that fascination when I booked the **Jack the Ripper walking tour** through the streets of **Whitechapel.** The tour was at night making it that much more unsettling, but this was going to be my chance to put the puzzle pieces together and solve the mystery for myself. If you have ever been to London's East End, Whitechapel even today is a pretty eerie place. It is still considered a poverty area, but one thing I noticed was that all the residents had cell phones, and satellite dishes in their apartments. But this was real history, and a chance to actually see the places I had only read and heard about. There I was standing in **Mitre Square,** where **Catherine Eddowes'** body was found mutilated 114 years earlier. Walking past George Yard where the first known Murder took place, down to **Miller's Court were Mary Kelly** was found a few months later in what has been described as the most gruesome murder to date. This is probably not a tour that is on most people's list, but history is history, and I have never been one to follow the crowd.

Did I mention this was a business trip, and I actually had a couple of meetings to address? Yes, we did have 2 days of meetings, but those are not very interesting to write about, and really are just a means to the adventure end. Since day 2 of the meetings ended early, it was time for myself and two colleagues to check out a traditional Oxford pub. Having been there before, I was fine with playing tourist guide, and having a bit of the English culture shared with my friends.

The next day it was home for what was a 5-day trip to England, that really should have been 3 days, but adding the week end to it, gave me that extra opportunity to try something new, and that's what adventure is all about.

Nice, France
August 2002

Having barely caught my breath from my first business trip overseas, I was advised by my leader that our division was going to be responsible for supporting a new technology platform in the USA. The platform was mostly overseas, but there were a few installations in the US, and we needed someone to have knowledge in case any support was needed. It was decided that I would represent my team, and I would need to travel to our Research and Development center located outside of **Nice, France.** The next training class was scheduled for early in August, less than 2 months away.

Since I had never been to Nice before (though Cannes is very close), this would be a new adventure. The company facility was located in the small town of **Sofia Antipolis,** but it was recommended that I stay at a hotel in Nice. Sofia was very small, and not really much to do in the evening, so I chose a hotel in the heart of Nice, right along the

Mediterranean. A very easy Air France connection through Paris had me arriving at the **Cote De Azur** in the late afternoon, as well as pumping up my Delta miles. The rental car process went smooth as well, and before I knew it, I was driving something called a **Twingo**, made by Renault. Everything was moving like clockwork the way one would wish all trips transpired.

The next couple of days would be spent in a training class, and the evenings trying various French restaurants. I was amazed at how smooth the traffic was, and how uncrowded a popular resort town was, but I certainly wasn't going to complain. I would find out on my next trip back to Nice as to why everything was going so swimmingly. I had scheduled a couple of days back in Paris after the training was over, just since it had been 12 years since I was last there and wanted to refresh my mind.

Because Paris is so large, I thought I would choose a named brand hotel, and I even fell for the hotel name that included "Central Paris" as part of its marquee. The hotel was not centrally located, and I had to take 2 Metro trains to get to the city center. Not a terrible inconvenience but a learning situation to be sure. I basically stopped at familiar haunts, the Eiffel Tower again, and of course my night time Seine River cruise. I also went back and explored the Louvre at a much slower pace this time. Taking my time, and really seeing the whole museum did not enhance my impression any more than the first time I was there.

One of the things that I liked to do was when I traveled on extended trips like this for work, I always tried to get my daughters and my wife a special little gift, so they could feel a part of my trips as well. For the girls I found these amazing music box carousels that were hand painted, and Mary got a French perfume that was only available in France at that time. My return home was straightforward on a nonstop Air France flight from Paris back to LAX and ended a very successful if somewhat short business trip overseas. One particular aspect of this trip that I did experience was a noticeable reluctance of French people to offer assistance for directions. Up to this point, I personally had never experienced the "so called" rudeness of the French people towards Americans. While I won't classify the experience as rudeness, it was clear there was a lack of friendly demeanor towards helping me find my bearings.

Nice France
JUNE 2003

As successful as my first Business trip to Nice was, it would be the polar opposite just 10 months later when I found myself back in the South of France for another training class. Even though the time span between trips was short, a whole lot of everything changed in that time. First, my company had the first of what would become regular re-organizations. After all the shaking was done, I found myself in the role of Implementation Manager, which was similar to what I had been doing, but with less customer exposure. It was not a choice I would have made, however in today's workplace, you need to be able to adapt, so now I was in Implementations.

Another change that took place was I shifted my traveling allegiance from Delta Airlines over to United Airlines. Delta had made changes in their loyalty program that made it more difficult to accrue mileage, and so I decided it was time to make changes in my airline partner. Since the actual travel benefits were for the most part eliminated, the only way to obtain travel perks was through the airline loyalty programs. I became known as what we call in the business travel industry a **Mileage Whore**. That basically means you go out of your way to stay with the same brand as much as possible, even if it is not the most convenient or efficient way to travel. I would soon find out this strategy had consequences as I prepared to head to France again.

I needed to travel to Nice again for a training session on an additional booking platform that my group would be supporting, and the person responsible for the training was in France. The easiest way for me to get there would have been to take Air France again, and accrue mileage on Delta, but since I was boycotting Delta I decided to take United Airlines instead. United did not fly into Nice, so I had to take a United flight to London, and then transfer to British Midland Airlines to get to Nice. Not only would I need to clear customs twice on this trip, but the change of airlines greatly increased the chances of lost luggage.

Those increased chances came to fruition as my luggage was lost and did not make it to Nice with me. I made it to the hotel but had no immediate change of clothes, and only had basic necessities from my carry-on bag. The luggage was located but would not arrive until the next night in Nice, and I had to be in the office the next morning. Sure enough, I had on shorts, and a shirt that had food stains on it, along with sneakers as I walked into the training class. Huge lesson learned about why you should travel with an emergency set of clothes in a carry-on bag, just as a precaution.

Again, staying in the city of Nice as opposed to the small town of Sofia Antipolis because of the night life proved this time to be a huge headache. My last trip to Nice was in August which I discovered was traditional vacation time in France, so the streets for the most part were empty. Now this is June, and boy did I discover what the Nice rush hour was like heading down the motorway. A trip that took 15 minutes the first time, now was taking almost 45 minutes with all the toll booths. The return trip was just as crazy as the traffic in the town Center of Nice was pretty insane.

Fortunately, my luggage finally arrived, and I finally had clean clothes as I had been in the same clothes for about 36 hours. The rest of the trip seemed to go pretty much as planned, and I had allowed myself the weekend to possibly make a trip down to Genoa Italy. I had not really been to the northern part of Italy and I thought this was a good way to add a couple of days of relaxation into a pretty harrowing trip. As I was preparing to take the train for that trip, the last shoe of this trip dropped, and it was one I could not ignore.

All through my career, traveling had never been an issue, but now I was married, and I had 3 little daughters at home. Daughters who were now old enough to know that dad has been gone for a while, and they were starting to miss him. I received a call from them and I guess there was some kind of disturbance in the neighborhood we lived in, where people were yelling and screaming out in the street. The girls heard this and got scared and called me crying telling me that they wanted me to come home. The feeling I had at that moment broke my heart, but I didn't know how I could get home any faster. I knew British Midland only had that one flight out of Nice, and it had already departed. A colleague of mine suggestion I call United and see if my ticket could be used on a partner airline to get home. The plan worked as the partner network was able to take my ticket and book me on a **Lufthansa Airlines** flight through Frankfurt Germany. I had to saddle up for a 13-hour Frankfurt to LAX flight in the back of the plane, but to make my kids happy, it was well worth the sacrifice. This was the first time that my home life would play a role in my needing to alter travel plans. I was becoming a responsible parent!

Russia (Россия)
September 2019 (Сентябрь 2019 год)

I honestly don't think there is a more misunderstood country from an American perspective than Russia. From the time I was a little kid, I was indoctrinated that the **USSR** was a sworn enemy of our country, and its way of life. It was not something that my parents drilled into my mind, but what our everyday media, and education taught us. Indeed, the political system that governed the vast country of what we conveniently called Russia, was in stark contrast to what we felt was the right system of government for people. The **Russian Revolution** of 1917 brought to power a group of individuals who believed that the workers should lead the country, as opposed to the Monarchy, and elite class. This thought believed that everyone should have the same, and that no one should have more than another. This kind of thinking flies in the face of those who believe that a free society should allow each individual to seek out their own fortunes, rather than having the State take care of everything in a person's life. That system was in place for over 70 years, until circumstances happened, and the people decided that another revolution was due. This time the revolution was essentially bloodless, and the Communist rule of the Soviet Union collapsed. Yet almost 30 years after that world changing event happened, there are those in the Western World who still view Russia from the Communist perspective. It is for that reason that I decided a trip to Russia was for me, and I would experience for myself the country that Winston Churchill once described as a "riddle, wrapped in a mystery, inside an enigma".

My fascination with Russia can be traced back to my high school days, and continued on into college through several Russian History courses. I always knew that the historical experience would be of interest to me, but the situation never presented itself for me to travel there. Even as I was planning a new adventure for 2019, traveling to Russia was not my first choice. My focus early on was to visit the **Holy Land of Israel, and Egypt**, but because the political instability and the constant fighting in that region, I was just not comfortable making the trip right now. I then turned my attention to making a trip to Iceland, which seems to be the "en vogue" place to travel to. The consensus among the travel community is that Iceland has become extremely expensive, and its relative remoteness did not allow me to take advantage of any of my travel loyalty programs to reduce the costs.

So that led me to look to Russia for my next life adventure, and it seems like all the puzzle pieces began to fall into place. I was able to secure hotel nights for free using the points from my Marriott account. Having free hotel space is such a huge incentive in making a trip financially feasible, and it made my trip that much more possible. Having the components all set, the only real issue I had to deal with was obtaining the necessary visa to enter the country. The process was an over the top application, and the most expensive counselor fees I have ever paid to travel. I had gone too far to back out of the trip because of expensive visa fees, so I just had to grin and bear it.

I chose to travel in September, hoping to avoid the busy Summer season, while enjoying what I hoped would be mild Fall weather. Transportation to Russia from the USA can be a challenge since there are no American carriers that travel directly into the country. That wasn't so much an issue, because like I did two years earlier on a trip to China, I chose to travel on a foreign carrier. My experience with International travel has led me to feel that American carriers are not nearly as good as carriers from other countries. This strategy served me well when I have traveled to Asia, but I would soon find out it is not a universal standard.

For this overseas adventure, I chose to travel on **KLM (Royal Dutch Airlines)**. In all of my years in this industry, KLM has always had a high reputation for service, and I was excited to have the chance to try them out. One of my flights from Amsterdam to Moscow would actually be using **Aeroflot** which is the Russian flag carrier. Visions of the old Soviet style airliners entered into my mind, but since the fall of communism, the state airline has starting using Western aircraft, and in this case the plane was a French **Airbus**, so I felt a little better. My flight to Amsterdam was certainly nothing to brag about, and truth be told, I found KLM's service to be very lacking, and their equipment to be old and dingy. As they celebrate their 100th year of service this year, the 747 I was on looked and felt like it was part of the inaugural flight. It was an experience that did nothing to redeem itself on my return flight home.

The timing of my flights could not have worked out any better as my flight arrived in Moscow at about 6pm local time. Once I cleared immigration and customs, which by the way was a breeze, and very much a surprise, I had a car service waiting for me. By the time I arrived at the hotel it was about 7:30pm, and being exhausted from almost a full 24 hours of travel, I went straight to bed for a good night's sleep. This allowed for a good rest, and the ability to start my first full day in Russia bright and early.

Moscow (г. Москва)

My first day had on the agenda a substantial amount of time checking out **Red Square**, and the surrounding areas. During my pre-travel research, I was already aware that the Kremlin is closed on Thursdays, so I planned several other things to do for this first day. Besides checking out the outer walls of the **Kremlin, St. Basil's Cathedral, GUM Department store**, and The **Russian State Museum**, I also had an introductory walking city tour booked, that was actually free of charge. The company operates this tour in the hopes that you will book other paid tours, but this 2-hour introductory was perfect for me. I was also hoping to see the tomb of Lenin on this day, but the waiting time was too long, and conflicted with this tour, so I was not able to see him on this day, but I would try again later. I also had an evening tour booked that focused on the important buildings and locations of the Soviet era. This tour took me to **Revolution Square**, the headquarters of the now defunct **KGB**, and many monuments honoring **Karl Marx,** and other Bolshevik revolutionaries. From a historical standpoint, I found the tour incredibly interesting, especially seeing how far Russia has come since that period ended.

I ended what would be a long and exhausting day with a night visit back in Red Square to see all the famous monuments lit up in the night sky. What I would discover throughout this whole trip is that Russia illuminates their cities like no other place I have ever seen. I suppose people who pride themselves on being "green" would not be too happy with the amount of electricity used to light all these buildings, but it was impressive nonetheless.

Red Square at Night, Moscow, Russia
September, 2019

My 2nd day would be all about the Kremlin, or so I thought. After a brief 30-minute introductory tour of the origins of the Kremlin, we were free to enter the grounds. The word Kremlin simply means fortress around a city, which is exactly what it is. There are numerous buildings and cathedrals within the walls of the Kremlin, as well as currently functioning government buildings where daily Russian governing takes place. The lines to enter the cathedrals were a bit long, so I opted to enter the Armory Chamber Museum first. I had read, and was told that this was a must visit, as it contained many items that were centuries old, and well worth the $15.00 admission price. There were some very impressive exhibits, but pictures were not allowed to be taken at all, and I really wasn't overly impressed, especially with the admission price that is in addition to the cost of entering the Kremlin grounds.

Since I ended up spending less lime in the Kremlin, and I had a lot of time before a special event I was doing later that night, I decided to do a little tourist shopping in the Arbat Street section of the city. Souvenir shopping at its finest, I was able to pick up a few things, and see the commercial side of what Russia has become. A couple of hours there, and a quick metro ride back to the hotel had me resting a bit for what was going to be a first for me, and a must do in Russia.

The Kalashnikov

Regardless of how one may feel about guns, there is no doubt that the number one gun of choice in the world is currently the Kalashnikov, or more commonly known as the AK-47. This is a Russian made piece of hardware, and is sold all over the world. A very popular tour in Moscow is to experience shooting these guns in a safe and proper setting, and now it was my turn to experience this thrill. I did a lot of research into the different tours, and selected a company called Moscow Gun Tours. I was really impressed with what I read about this company, and the fact that the Marksman, while Russian, was raised in Canada, and spoke fluent English. Given the nature of weapons and safety, I felt a strong English background was a good selling point. I contacted the company and explained about my trip and interest in shooting, and almost immediately I received a response from Evgeny, the owner of the company. He confirmed the date, time and my hotel location right away, and asked about my agenda while in Russia.

On the day of the shooting, Evgeny met me at my hotel lobby, and we walked to the closest Metro station where we caught a Metro train to the shooting club. Once all the paperwork was done, he took me into the shooting area, showed me all the safety aspects and functions of the gun, and then let me do my thing. It was a great experience, and while I had shot a gun many years before with one of my dad's hunting rifles, I was by no means an expert. The kick of the AK-47 was more than I thought, but I quickly adapted, and had a wonderful time trying something new, and unique. After the session was over, Evgeny made sure I got back to my hotel, and I called it a night. I appreciated his hospitality, and professionalism, but I would soon find out the extent of his kindness before my trip to Moscow ended.

Evgeny from Moscow Gun Tours, Moscow, Russia
September, 2019

I took a day to experience some of the Russian countryside by visiting two very old cities, *Vladimir & Suzdal.* Both of these cities date all the way back to the year 862, but while Vladimir has grown into a major city, Suzdal has retained much of the old Russian charm, though it has succumbed to what we might call a tourist trap. It was definitely an old Russian small town, but it was also filled with many souvenir shops and fee-based museums, capitalizing on the tourist trade. After spending a couple of hours there, I took the bus back to the town of Vladimir in hopes of seeing something less touristy. By this time, I was starting to feel a bit Kremlin'd and cathedral'd out, and really just wanted to experience ever day Russian life. After walking a bit, I came across an English Beer Pub, and I knew this was just the place. The people including the bartender spoke almost no English, but we managed to have a good time for a couple of hours, and they seemed genuinely intrigued by me being an American. Even though English was scarce, the music system was playing all American music as evidenced by The Doors Greatest Hits in its entirety. A final train ride back to Moscow from Vladimir, and this 14-hour day was in the books.

For my final day in Moscow, I didn't really have an agenda, and was probably just going to kick around. But then I was contacted by Evgeny, who wanted me to experience his city, the way a non-tourist would experience it. He met me at my hotel, and he spent the day showing me many things I would have never seen without his knowledge. We checked out Lenin's tomb, as I was unable to see it on my first day because of the crowds. This time we were first in line, and went through the mausoleum in a quick manner. I have to say, I was extremely moved having seen one of the most important world historical figures of the 20th century, and it would be one of the top highlights of this whole trip.

A visit to a local flea market type place for some last-minute souvenir shopping, checking out the space monuments, and the exposition park rounded out this non-tourist sightseeing day for me. I was very grateful for the time that Evgeny volunteered to help complete my trip, and the only payment he would accept was for me to share my experience in his country, and to tell the American people that Russia is a wonderful place to see, and Americans are encouraged to visit. Thanks to Evgeny, not only did I have the full Moscow experience, but I also have a new global friend that I will add to my list of amazing people that I have met through my travels.

Saint Petersburg (Санкт-Петербург)

The second major city that most people know about in Russia is Saint Petersburg. While it is not the size of Moscow, it is nonetheless a large city based on any standard of measurement. Saint Petersburg has a much more varied history than Moscow does in that it has gone through several name changes, and has seen the brunt of a World War, that would have left many a city in ruins. Founded in the early 1700's by Peter the Great, Saint Petersburg is not nearly as old as Moscow, but still has a long history filled with royalty given that it served as the Capitol of Russia for about two centuries. During the turbulent times of 1917, the city's name was changed to **Petrograd,** after Czar Nicholas II was overthrown by the provisional government. A few months later, the Bolsheviks overthrew the provisional government, and shortly after that the name Petrograd, was changed to **Leningrad**, and remained that way for 70 years. With all this history to be seen, it's no surprise that I was really stoked to visit the second largest city in Russia.

As with many of my travels, particularly in Europe, I decided to take the high-speed train from Moscow to Saint Petersburg. 27 years earlier on a trip to Scandinavia, I had an opportunity to take a week end excursion to what was then Leningrad. Traveling with a new wife, who did not share the same adventurous lifestyle that I did, we decided to pass on that part of the trip. I really didn't know if I would ever have another chance to see that city. But as fate had it, here I was on a brand-new trip, living a life I did not envision all those years ago, traveling to that very different city. As with all things in Russia that have improved with the fall of the Soviet Union, train travel has become just as sophisticated as the rest of Europe. So, a trip that used to take 8 hours has now been cut in half with the new high speed Sapsan trains, and has even tossed in first class service.

Arriving in Saint Petersburg in the early afternoon was a good idea, but I still needed the weather to cooperate to follow through with my plan. I left the day open from touring just so I could get my bearings and find my hotel. I did a little walking in a light rain, and was able to learn the route for the events I had planned for the next day. After a brief rest, I had a night cruise of the Neva river, and the opening of the draw bridges planned. I couldn't figure out why this tour left at 12:30am, but soon learned that the bridges are pulled up at 1:00am every day and left that way for 4 hours to allow all commercial shipping to exit and enter the city. Seeing all the buildings and palaces lit up at night was incredible, and toss in a little Jazz music on the boat, and it made for a great start to my visit, even at 2:00am. What I truly learned was that Saint Petersburg is an incredibly beautiful city, especially at night.

Neva River at night, St. Petersburg, Russia
September 2019

The next day I had a morning city tour planned, that would take me through many of the main tourist points of the city, including Palace Square. What made this tour important was that it was a free tour, sponsored by the tourist bureau. Much like the tour that I took in Moscow, here I was on another tour free of charge, with the only requirement of tipping the guide if you felt they did a good job. The main point of this is that there are many opportunities that exist like this, if you are willing to do the research, instead of booking the first tour you find. Under normal circumstances, a tour like this would have set me back about 100.00, but a little digging, and I was able to enjoy this for a 10.00 tip.

The weather was off and on, but we managed to make it through without too much rain. After a couple of hours of walking and learning about the history of St. Petersburg, it was time for something a bit different. Russia is well known for vodka, although I think much of it is overblown hype, it is a popular drink in the country. So, I decided to check out **the Russian Vodka Museum,** and learn a bit more about the national drink. As I had found with most museums, the entrance fee was a modest 7.00 USD, so it was well worth checking out. They offered an audio guided history of how vodka came to be, and then at the end, the opportunity to taste 3 different vodka types. The tasting included a traditional Russian snack that consisted of white fish, pork fat with bread, and a salted cucumber. The salted cucumber is more commonly known as a pickle. It was interesting to try, but the white fish and pork fat had me reaching for the 3 vodka shots faster than I thought was possible. In the end, it was worth the experience, and a bonus to try a tiny bit of Russian cuisine.

Long on my list of travel adventures has been to see the **Hermitage Museum**. Listed as the second largest museum in the world next to the **Louvre** in Paris, the Hermitage has over 3 million exhibits displayed across its several buildings. Located in Palace Square, the main building is the **Winter Palace,** and it houses a variety of paintings, sculptures, and many Russian historical artifacts. The experience of seeing all this art was amazing in itself, but add to that the buildings were largely destroyed in World War II, and you really have an architectural experience that to me was 2nd to none. My Hermitage experience lasted about 4 hours, but could have been so much longer if I had tried to see a lot more, I just became somewhat fatigued from the all of the walking. The Hermitage, just like most of the other museums I visited, boasted a very favorable 700 Ruble entrance fee. That is less than 10.00USD. Imagine if Disneyland offered a 10.00 entrance fee, and the chance to actually learn something while you were there. The point of all the entrance fees to these famous museums is for all people to be able to afford to experience them, something we sorely lack in America.

The Winter Palace & Hermitage Museum, St. Petersburg, Russia
September, 2019

The life of royalty is something that most of us never experience or understand, so it was not surprising that the day after I spent time at the Winter Palace, it was time to take a trip to the Summer Palace, better known as **Peterhof Palace**. Peterhof is located about 45 minutes outside the city center, and the fastest way to reach it is by hydrofoil, or high-speed boat. So, I booked a ticket on the boat, and jetted out to the Palace. The weather turned out to be beautiful, but a bit cold, which was good because this was a total outdoor adventure. The most impressive thing about the palace is that there are over 150 fountains on the grounds, and none of them use electric pumps. All the fountains are driven by gravity, and require no motors or electricity at all. Sadly because of the preparations for the final days of the fountain's seasonal operations, many of the fountains were not turned on, but the grounds were impressive nonetheless. A couple of hours walking around the grounds, and watching a few musical presentations, and I felt like I had seen what the Summer Palace was all about. An afternoon boat ride back had me feeling a little tired, and thinking about what I still wanted to see. I went on a pub search close to my hotel that night, and found a small little basement bar called The Beer House. After 9 days in Russia, I found myself longing for some American food and language, and this place had it, so I checked it out for dinner. The place did not disappoint, and had an unusual amount of foreign beers on tap. With so many to choose from, I gave an Austrian beer called Edelweiss a try, and in fact I tried a couple of them. A successful day nine was in the books, but what to do on my final day?

Peterhof Palace
September, 2019

When I started to map out this trip, I knew I had one day where nothing was planned as I wanted to allow for any shifts in the itinerary in case things changed. One option I thought about was taking a train to the city of Vyborg. Located near the Finnish border, Vyborg is probably like many other of the cities I encountered. Great architecture, and loads of history were probably the bulk of it, but I was feeling a bit tired and opted not to do another all-day sightseeing event. I decided that I would spend my final day taking in the cathedrals in depth, and just finalizing my souvenir commitments. I took a walk over to the **Church of The Savior on Spilled Blood**, where Csar **Alexander** II was assassinated, and spend a little bit of time there. The first day I was there, it was raining a little and the overall weather wasn't the best, so I probably rushed through my time there. Next, I was going to head over to **Saint Isaac's Cathedral**, and had a chance to climb the colonnade, and see the city from the highest point. The views of the city were amazing, but the 226 step climb through the narrow stairway was interesting to say the least. My fear of heights climbing the last stair case kicked in, but I gutted it out, and was able to take some incredible pictures of the city.

My final night of the trip was pretty low key since I had a very early flight the next morning back to America. I had to leave for the airport at 3am, so I had an early dinner, and a last walk around some of the canals close to my hotel. It gave me the opportunity to really think about what I had seen the past 10 days, and just how much our own environment molds our perceptions about places we have never been to. All of the perceptions and views that our media and news have spewed for a generation, have been nothing but a huge lie designed to promote a misguided agenda. So, if you still had any lingering questions, let me assure you that the Soviet Union is no more, and the Russian people are warm and friendly, and would love to have Americans visit their country. My own personal take away is that I have had the opportunity to live out all the things I read in those history books, and no amount of money can purchase that.

Final Thoughts

No trip that I have taken to date has elicited the question of "why" more than Russia. Although the old Soviet Union has been gone for almost 30 years, it is still the most misunderstood country among the masses in America. Most of the stereotypes that we grew up with still prevail, much to our own detriment. Because we are a society that relies on others telling us what to think and feel, we have missed out on the opportunity to discover for ourselves what is really true. In the case of Russia, no amount of CNN or social media can accurately portray the experience I lived for 10 days.

What's Left?

Having pretty much covered what is referred to as Western Europe, there isn't a whole lot left that I have to see. Oh, there are some places I would be happy to go if life takes me there, such as Spain, or Portugal, but I don't see myself making a special trip there for myself. I do however plan to see Greece, with the Greek Isles, and parts of Turkey, particularly the area known as Troy

Even though I have satisfied a requirement to experience Russia, I would still like to travel on the Tran Siberian Railway through Russia all the way to China. While I visited the two major areas of Russia, the country is so vast, that there is so much more I would like to see. There are still many parts of Eastern Europe that I have not seen, but on the whole, I believe I have seen a great deal of the continent that many American ancestors came from.

It's hard to put into words what I have learned from my experiences in Europe, but the feeling of being able to see different cultures, and see the similarities, along with communicating in different languages, takes the concept of travel to a whole different level

5-Asia

When you think about visiting the largest continent in the world, the thing that comes to mind is where to go. There are so many cultures the encompass this mass continent that most of us often forget about the many different countries and cultures that make up Asia. There is also a mysteriousness about Asia, and its history which dates back thousands of years. Probably my very favorite class in high school was when I was in 10th grade, and the class was World Cultures. One of the assigned text books we read from was called **China, Japan, and Korea.** It was a book that we would read when studying that part of the world in the class. I was so impressed with that book, that I knew I had to see the things I read about. Asia not only includes the traditional countries that we often think of like China, Japan and Korea, but it also includes countries who are often geographically labeled as Middle Eastern and India. With so much land to cover, I knew it wasn't a matter of "if" I went to Asia, but "when" I would finally get there. And get there I have in some of my most adventurous travels to date.

South Korea
AUGUST 2007

It seems hard to fathom that I would be on the precipice of making my first trip to Asia, and yet be in a position to where I absolutely did not want to go. That's where I found myself in late Summer of 2007, and it was a difficult place to be. It had been about 5 years since I made my last international trip for business, but in those 5 years my life shifted quite a bit. I was now working with the Sales team in securing new business, a position that I really liked by the way, but one that had increasing pressures to perform. As a family, we had moved into our "Dream Home", at a time when the market was on the verge of collapsing out from under us. I also started to feel that my marriage was showing the cracks that had been underneath, only now they were becoming more visible. Finally, the first major fundamental shift of my life had now just occurred with the sudden passing of my friend Raj Prasad. Raj hired me into the role at my company, and I considered him a huge mentor in my career, and suddenly one Thursday morning, he collapsed at home, and was gone.

While all of these things were going on, I was chasing the biggest fish of my sales career, and had them biting at our company hook. This potential customer was the kind of deal that can turn you into a real company hero overnight, not because of the size, but because of brand recognition. So, during all of this turmoil, and in the midst of funeral preparations for Raj, **Samsung** had requested a meeting and presentation in **Seoul, South Korea.** I remember vividly, as I was sitting in Business Class on that **Asiana Airlines** 747, how unprepared I felt, and how unmotivated I was to even be going. The term "the show must go on" was particularly applicable here, and not even the death of a friend could stop it.

Buckled in for the twelve-and-a-half-hour flight to Seoul, I had a lot of time to think about things, and about this presentation. I knew my motivation was low, and it started right when the flight attendants brought the first meal of the flight. I love Asian food, but I wasn't up for trying the Korean delicacy, and settled for the rubber chicken dinner as a safety net. Upon arriving at **Incheon Airport**, a transfer was waiting for us, and took us directly to the hotel. When we arrived at the **Shilla** hotel, things were looking much more promising. This was a 5-star hotel that was owned by Samsung, and it did not disappoint. The Asian culture is most notable for service, especially in the hospitality industry, and that tradition was evident from the moment we walked into the lobby.

The first day of meetings with Samsung went well, and the fact that Raj was a big part of this engagement was not lost on the Samsung team, as each of them expressed their condolences about his sudden loss. We were their guests, and they were going to do everything possible to ensure we were well cared for. They took us out for dinner that first night to a traditional Korean style meal. The process of a Korean meal is a bit different in that side dishes are served in tiny individual dishes. The main course is usually brought out in the center of the table and is cooked on one of the small Korean grills right at the table. Things started to get a bit dicey for me, as the main dish is a community help yourself, where you stick your chopsticks right into the main course and help yourself. That is where "East meets West" came into focus for me, as I watched chopsticks go from individual's mouths, into the main course. Once I got over that shock, it was time for the soup to be brought out. It was a seafood soup that contained octopus, including the little suction cup tentacles. As one floated to the top, I was suddenly reminded of the old Three Stooges Shorts where Curly would always fight with his soup at the dinner table, and it would reach up and pinch his nose. Somehow, I made it through the culturally shocking dinner, but I found out that with Samsung, the party was just getting started.

One element of working in Sales is when a customer wants to entertain you, you must ingratiate yourself, and go along with it. So after dinner, they took us to another club where I would get the chance to experience the Korean version of the Kamikaze drink called a **Soju Bomb.** These are done as shots, and the affect is not instant, so they kept giving them to me. Needless to say, after an hour, I had to be escorted back to the hotel, as I could barely walk. I also had the fear that I would not be able to wake up for the meeting the next morning. I did manage to wake up with the biggest headache possible, along with a sour stomach. I made it through that day of meetings and presentations but made sure that our evening was planned for an early night. Though I felt pretty much like crap that 2nd day, I knew I had impressed the Samsung team, and really had a great cultural experience.

I had allowed for one day of some sightseeing and souvenir shopping as my commitment to my girls about overseas gifts was still in force. I was able to check out the Seoul Museum and learn much more about the Japanese occupation of the Korean Peninsula prior to World War II, and just how much that remains a touchy subject in Korea today. Also, being only 30 miles from the DMZ, and the country of North Korea, had a surreal feeling all its own. A little shopping in the market district of Seoul netted me some authentic Korean gifts for the girls, and a small temple bell statue that rings for peace and tranquility for myself.

My first experience with Buddhist Culture in Korea, August 2007

A bit of luck occurred on our way home as Asiana overbooked Business Class, and my colleague and I were upgraded to First Class. Silk pajamas, flat laying seats, and amenity kits were provided before take-off, so I knew the flight home would be peaceful. Though I felt the trip was a success, and I had finally made it to Continent number 5, I could not help but think of all the issues still waiting for me back home. It was a defining moment in my life, but I also knew that I would return to Asia again, as there was so much to still see in the World's largest continent.

China, Tibet & Nepal
SEPTEMBER/OCTOBER 2017
(MY MAGNUM OPUS)

It would be 10 years before I would realize a dream and make a return to Asia, but those 10 years contained many life changing moments. I still continued a heavy business travel load, but the vacation travels stopped as I found myself traveling down a lonely and unfamiliar path. During that span of time, I would deal with death, accidents, more death, divorce, a broken engagement, another death, more death, and finally 2 more deaths to finalize that 10-year gap. But I also managed a few good things in that span as well as, I discovered a love for writing that would not only help me deal with all the loss, it helped shaped an alternative outlet for me to seek. Publishing 2 books in that time span was for me an unexpected milestone that I would achieve, to combat all the difficult challenges life tossed at me.

Having come through all of that turmoil, people started to tell me that I needed to take care of myself and do something for me. The girls were older now, as Megan was reaching adult age, and she would be my last child to reach that plateau. I started thinking about taking a trip just for me, one that I could go and do what I wanted to do. It had been 28 years since I had taken a vacation by myself, and I knew it was time.

I started thinking about where I could go, and what I wanted to do. I had a significant number of American Airlines miles accumulated and some hotel points that would net me some free rooms, so I just needed to map out where I truly wanted to go. I had never been to Africa, and if I were to go, it would have me visiting continent number 6 with only one left to go. Then I started thinking about Asia, as my last trip there was so short, and I really wanted to explore it in much more depth. China has always been on my must-see list, and I thought this would be as good of a time as any to do it. I always wanted to see the **Great Wall,** and some other places, but since I was going that far, I also wanted to see something that not everyone has the chance to or has the desire to. That had me thinking **Tibet**, and the mysteriousness surrounding a closed culture that was just very recently opened. I had made the decision, and now I started a journey that would have me the most prepared I had ever been when taking a vacation.

There was much to prepare for with a trip of this magnitude, the first of which was to secure a Chinese Visa. I needed to have a Chinese Visa in order to apply for the **Tibet Travel Permit** and to book the Tibet portion of the tour. Special permission must be granted to enter Tibet, and you must be escorted on some kind of tour in order to travel there. You just don't show up in Tibet, and say, "hey I'm here". The political stability in Tibet is always heightened as the Chinese maintain strict control over every aspect of the Tibetan lifestyle.

The trip expands

I had my China flights all confirmed, the hotels in Shanghai and Beijing were confirmed, and I had the plan to travel to Tibet, but it still didn't seem grandiose enough. After all, this was my magnum opus, the trip of all trips for me, so I needed something a bit more. It never occurred to me that half of **Mt. Everest** is in Tibet, with the other half in the country of Nepal. Suddenly, a trek to stand at the **top of the world,** seemed like the ticket I was missing. I was able to find a tour that included a 2-day visit to **Everest Base Camp,** and I immediately booked that. Since I was so close to Nepal, I figured why not take a couple of days in **Kathmandu,** and experience more of a Hindu culture, along with the Buddhist culture of Tibet. There you have it, a 19-day trip to China, Tibet and Nepal, that would see me do many adventurous and bucket list things.

The months passed quickly, and I had all of my tours and transfers planned. I had been to the doctor and had a physical to ensure my health was good to make this kind of high altitude trip. I also secured some medication to help alleviate the effects of high altitude sickness. This was probably the number one concern I had about taking this trip. For 2 weeks I would be above 3000 meters, and above 5000 meters for one of those weeks. There wasn't a stone left unturned that I could think of, because I wanted this trip to go off with as few hiccups as possible. As I found out the strategy would pay off, and while there are always unexpected things that happen, they were very few and far between.

Before I knew it, September 21st was here, and it was time to depart. I had chosen to use my American Airlines miles on partner carrier **Cathay Pacific**. I had always heard that Cathay was one of the best airlines in the world for service and decided this would be something I wanted to experience. Flights on American were more direct, as Cathay had me transiting Hong Kong, but to me the extra travel time was worth it, and I made the most of the brief time in Hong Kong. The flight to Hong Kong was my longest flight to date, a mere 15 hours nonstop, but as predicted the service was top notch, even in Coach Class. A change of planes had me heading for my first destination, the largest city in the world by population, **Shanghai.**

On my arrival in Shanghai, I was met by a young lady from the transfer company, and a torrential rain storm. She helped get me to the hotel and advised me about my itinerary for the next day, which was a sightseeing tour of Shanghai, and one of the Water Towns nearby. The next day it was raining again, but I had to make the most of the day, as this was the only full day I had in Shanghai. We visited a Water Town, which is basically a village along one of the rivers, where we would find interesting souvenirs, and small business owners making a living. The goal was to also check out the financial district of Shanghai, along with the Bund, which is the marketplace, but the weather deteriorated, and the financial tower was shrouded in fog, so nothing was visible. All in all, not a good way to start a trip, but these are things you just cannot control. Hopefully it would get better.

The next morning, I caught a train to the imperial city of Beijing, which was about a 4 ½ hour ride via high speed train. The train was modern, complete with Wi-Fi, and all the possible outlets you could ask for. Beijing is really where the history of China is, and while Shanghai is bigger, I felt Beijing had so much more to see. When I arrived in the early afternoon, I had a taxi take me to my hotel. I had a recommendation to stay at the **Crowne Plaza Wangfujing,** and man was that spot on. The hotel was located right in the heart of most everything, and soon I was walking down to what would be the **Forbidden City, and Tiananmen Square.** I could not believe I had made it to one of the most historic, and infamous places in the world, but there I was standing at the gates. The next day would be a full day tour of this area, along with a stop at the **Temple of Heaven**, and the **Summer Palace.**

The day tour around Beijing was really fun, and I met some really good people on the tour from all over the world. Tim and Karyn from America, and Viktor from Hungary (who would be joining a different tour to Tibet at the same time as me). It was a long day of touring, but the guide was great, his name was Peter, as all Chinese tour guides have an American name, and many have the name Peter. Probably the highlight for me was having my picture taken with a **Beijing Police Officer**. Most of the ones I asked refused rather abruptly, but one young man at the Summer Palace was very happy to pose with me. I hope he did not disappear from civilization as technically that is a "no no", posing for pictures. Tim and Karyn were telling me about what awaited me the next day, as they did their tour of the Great Wall the day prior, and did exactly what I signed up for.

Picture with a Beijing Police Officer, September 2017

It was here, the first of a long-time dream to see the awesome **Great Wall of China.** Now most people who visit the Great Wall, book a tour that will visit the refurbished area of the wall. There is nothing wrong with that because it is an amazing experience, and still very challenging, but I am not most people. I booked a tour that started with a hike to an unrestored area of the wall and continued on to the restored section. It would end up being a 10 Kilometer trek, seeing both the wall in its original state, and its restored section. The tour picked me up at my hotel, and the tour host Peter (not kidding), loaded me into the van, and away we went to the start of the hike. It would take about 45 minutes to get to this remote section, and there were a few detours along the way. Finally, we arrived at the starting place for the hike which was a wooded area with a narrow trail. The ascent would be about 500 meters in grade, and about 2 kilometers in distance We started to walk the trail, and it was your basic hike on a trail, but about 10 minutes in, the ascent up the mountain would start. The weather was overcast and misty, but it was also very humid. We could see parts of the old wall as we were ascending up, and the watch tower that we were heading for was becoming clearer in focus. About 2/3 of the way up, the humidity was starting to take its toll on me, and I became very dizzy at one of the rest breaks. I started to panic because we were in the middle of nowhere, and I wasn't sure I could make it up the mountain. Fortunately, one of the ladies' hiking with us had some chocolate, and she gave that to me. I ate a couple of pieces, found a rock to lean against, and put my head back and closed my eyes. So much was running through my head at that moment, I had failed on the first significant challenge, how would I get out of there, would I not see the actual wall? After lying still for about 10 minutes, the dizziness subsided, and I felt like I could continue. The guide Peter stayed with me and made sure to walk behind me the rest of the way.

I pushed through and sure enough we arrived at the watchtower for what would be the start of the hike on the wall. This area is only open to hikers, and not the general tourist. It was a section of the wall that was in its original state and had not been restored in anyway. Many elements were broken, and in some cases dangerous to walk on, so we had to be careful. Much of this area was covered in brush, and considerable weed growth, and many of the pathways were very narrow. After about an hour of hiking, we reached the highest point of the Great Wall, and entered into the restored section called **Mutainyu.** Here the bricks, and the walkways were completely re-done, and the watchtowers were great viewpoints. Even though this is the restored section, the climb and the terrain were still fairly difficult. There were numerous steps to walk, and steep grades, but here I was hiking the Great Wall of China. The end point for me came at Watchtower 6, where I would take an alpine slide ride down the mountain, as opposed to walking down hundreds of steps. Altogether, the hike was 10 kilometers long, and my legs and feet felt every inch of it. It was certainly worth every stiff joint I would endure, for a chance to see one of the greatest man-made structures in the history of our world.

For my final day in Beijing, I had contemplated the idea of going out to the Beijing Zoo to see some panda bears, but upon waking up, my body felt so stiff from the day before, I decided against it. I spent a little more time shopping and one more visit to Tiananmen Square before heading back to the hotel and checking out. I was on my way to the **Beijing West** railway station for what would be the longest train ride I had ever taken. If there was one thing I will never forget about Beijing, it was the look on the cab driver's face, when I paid my fare. 100 Yuan is approximately 15.00 USD, my fare came to about 7.00 USD, and I gave him the 100 Yuan, and told him to keep the change. He almost broke into tears, as that money goes a long way in China. It really put things into perspective for me when thinking about the things I have in my life.

Train number 21 would leave Beijing at 8:25pm on Thursday night, and arrive at its destination in Lhasa, Tibet at 12:30pm Saturday. So, it was time to settle in for the long haul that would include sleeping 2 nights on a small couch in my room, with 3 other people. The long-distance trains are the older, and much slower cars than what they use for the high-speed routes of Beijing and Shanghai. There was not a whole lot of spectacular scenery to see along the way, but rather the economic displacement between rural and urban China. It was mostly farm lands, and small towns, with an occasional city along the way. By the time we came to the town of **Xianning**, the landscape had changed, and so did the altitude. The train attendants passed out a little card stating that we would be going above 3000 meters, and that you must be in good health to withstand that altitude. Oxygen was being pumped into the cars for the rest of the journey.

Late on Friday afternoon we would arrive in **Golmud.** This is where the **Qinghai-Tibet** train takes over for the rest of the voyage. Reaching altitudes at over 5000 meters in height, this is the highest railway in the world, and is built above the permafrost. This part of the railway was completed in 2006 and became the alternative way to travel to Tibet. By this time, I was feeling a bit train crazy as we still had about 18 hours of travel left, but we had now crossed into Tibet. I was surprised to find that when I woke up the next morning, there was no snowcapped mountains. I thought that being the beginning of October, the snow would have already started, but it was unseasonably warm, so the snow would have to wait. Somehow, we finally made it when the train pulled into Lhasa station, and the 42-hour adventure was complete.

The roof of the world

Tibet is often nicknamed **"the roof of the world"** because much of the country sits at an elevation of over 11,000 feet. By comparison, many people think of a mountain city like Denver as high altitude, but Denver sits at an altitude of 5200 feet above sea level. Lhasa, the capitol city of Tibet is over twice that high, so consequently altitude sickness can be a real issue if you are not careful. I had already started taking my medication to help with the altitude and was therefore feeling really good. When we exited the railway station, we immediately had to go to immigration when they saw my American passport. Another gentleman traveling on my tour, named Will, had a British passport, and there were some questions concerning his Tibet Travel Permit. After a few extra minutes, and a few scowling stares from the Chinese immigration people, Will was cleared, and we were met by the representative of **Tibet Vista Tours.** He presented us with our white Tibetan scarfs, and proceeded to tell us where we could go unaccompanied and were we could not go to. Suddenly, the notion of a suppressed culture came into focus, and going with the flow was definitely the strategy.

The hotel in **Lhasa** was in a great location, within an easy walk to the **Barkhor shopping area,** and the **Jokhang Temple,** which we would see the next day. While the architecture of Lhasa was one of older Tibet, the city itself was fairly modern with both local business, and Western influences as well. For the next 4 days, I would be immersed in the Buddhist culture, and visit a city that until recently had been closed off to foreign visitors.

The **Deprang Monastery, the Sera Monastery,** and Jokhang Temple were nice to see, but the real reason I wanted to see Lhasa was for the **Potala Palace.** That was the home of the Dalai Lama when he was the leader of Tibet until 1959. The highest residence in the world overlooks the City of Lhasa and is spectacularly lit up at night. The next morning, we took a tour of the Potala Palace, and many of its magnificent rooms. There are many rooms that are off limits to tourists, but we were able to see quite a bit of it, and it was amazing. For our final night in Lhasa, we were treated to a traditional Tibetan dinner and show, including dancers and exotic costumes.

Up to this point, I was feeling great with no issues of altitude or food related things. However, as we started heading West into the Tibetan plateau, I started to feel a bit squeamish, possibly because of the spicy soup I had the night before. It was OK because we had quite a long ride in the van and a few stops before we would arrive in **Shigatse** for the night. Our first stop was one where I got to check one of my must "see's" off of my checklist. Right before we arrived at **Yamdrok Lake,** we stopped at a mountain view area, where before my eyes, I saw them. **Tibetan Mastiff's,** the largest canine breed in the world. These dogs are massive and have a long mane like a lion. They were perched up on tables where you could have your picture taken with them, for a fee of course. The cost was like $1.50, so it was well worth having two of those big boys surrounding me. I also got to see and sit on top of my first Yak, as they are the main cattle in Tibet for food and fuel, more on that in a bit.

We arrived at Yamdrok Lake, and again stopped at a spot that was ideal for pictures, and then drove down to the lake shore for a 15-minute rest period. The further West we got, the more primitive Tibet became, especially when it came to restroom facilities. Squat pits had now given way to open pits, and the smell was beyond anything I could even describe, and you had to pay to use it. After a very long day of driving and sightseeing, we arrived in the city of Shigatse, which is the second largest city in Tibet. Very surprising to me was that it is fairly modern, and the hotel we stayed in for the evening was quite nice.

Reaching the top of the world

We had another long day of riding in the van, but today it would all be worth it as we headed toward what would be the pinnacle of this trip, and bucket list item number 2. The ride wasn't without its moments though as the Chinese closed a portion of the main road because they can, and we had to take a detour over an unpaved road for a total of 2 hours. After having my innards juggled for 2 hours, we reached the small town of **Tingri.** There were few food options available, but by this time, I had pretty much cut myself down to apples and bananas. The cleanliness factor of people not washing their hands, especially after using the bathroom, and working with food, was just too much for me to comprehend. After lunch we had this heinous checkpoint where we had to exit the van and produce our paperwork. It took over 90 minutes for us to get through, and by then my last nerve was pretty much gone.

We climbed the winding mountain road for about an hour and reached the top at about 5100 meters and pulled over to the view spot. There it was, the entire Himalayan mountain range straight out, and the granddaddy of them all in the middle, **Mt. Everest**. Suddenly a dream that started 10 months earlier, with hours of planning, continuous months of workouts, and even a few naysayers who felt it was too dangerous, was realized. Even though we were still 2 hours away from our destination, the view was simply stunning. Back in the van, for now we had to go down that winding road as we would head to our destination for the night, a primitive tent community at the Base Camp of Everest. We were warned about the primitive conditions a head of time, but still the experience is one that must be seen, or not depending on your standards. There were about 15 people to a tent, and there is no running water. The tents are heated with a stove that burns, get ready for it, Yak Dung!! It was only for 1 night, so I knew I had to grin and bear it.

Another scary reality struck as one of the young guys on our tour complained of a severe headache. This is the most common symptom of altitude sickness, and the tour guide immediately got him on oxygen. We were at 5200 meters, or approximately 17,100 feet above sea level, and that height can be deadly if you are not prepared. The oxygen seemed to help, and he was able to get some sleep, but he was not the same for the rest of the trip.

The plan was to wake up early and catch the sunrise over Everest, but we had to do some hiking to get as close as we were allowed. When morning came it was still pitch black and freezing cold in the tent. One of the caretakers came in and pulled out a big bucket of Yak dung, reached her bare hands into the bucket, and put several scoops into the stove. She then lit the fire and turned to me and asked what I would like for breakfast. I politely declined and continued to get dressed in my winter clothes and hiking boots as the temperature was under 20 degrees. This was it, this was the day to experience one of the 7 natural wonders of the world, and to stare at the top of the world.

As I started hiking toward the base camp, it was still dark out, but the moon was draped over the mountain, and very visible. I remember thinking about the movie **National Treasure** when Nicholas cage found the ship located in the Arctic ice, and I uttered his words **"hello beautiful"** when looking at Everest. All along the hike, hundreds of people were positioning their cameras and tripods for those few minutes when the sun came up over Everest and presented the appearance of multiple colors on the mountain. Just before the sun came up, I had made it to the milestone marker of base camp. I did it!! I stood before the highest point in the world, a place where more than 250 dead bodies remain stranded in the Death Zone. A challenge I presented myself with was achieved. I took many pictures of the moment, but the one I remember most was the picture I had taken with me holding a sign telling my daughters that "I did it, and they could too" with a sun-drenched Everest in the background. A moment of self-achievement that will forever be engrained in my memory, along with the gratitude that God allowed me to look upon this wondrous site.

My message to my daughters as I reached my goal, October 2017

As we started our 2-day journey back to Lhasa, all of the things we endured getting to Everest, would be revisited on the return. The long unpaved road, the annoying checkpoints, and the small-town eateries where washing your hands is the unwritten sin. We managed to make it back to Shigatse, again late in the evening, and exhausted. Our final day of the tour consisted of one last visit to a monastery in Shigatse, as if the previous 5 on this tour were not enough. One saving grace was that I met some wonderful little Tibetan children here, and they were kind enough to take pictures with me.

Tibetan Children, October 2017

We had about a 4-hour drive before we would get back to Lhasa, so I plugged in my ear phones, and let the iP-hone do its thing. We got back to the hotel in Lhasa around 6pm for my final night in Tibet. By this time, I was so over Tibetan food, or lack of it, that I headed straight for the **Burger King** in Barkhor Square and ordered a Double

Whopper with cheese. My first American style food in 2 weeks, and it tasted pretty darn good. As I crawled into bed that final night, I remember feeling physically exhausted, and my body sore from what I had put it through the previous 2 weeks. I quietly remember thinking that I wished I was heading home the next day. But this was only the completion of part 2 of a 3-part adventure, so I still had one more piece to this travel puzzle. My 2 weeks in China were completed, and it was one of the most well put together programs I had ever experienced. Having expected to have some difficulties, I can't say enough how impressed I was with Tibet Vista Tours, and China Highlights for the meticulously prepared itineraries I had.

Thank You Bob Seger

Once I reached this point in the journey, I was originally scheduled to fly to meet my niece Molly in Chongqing, who was teaching English there. I would spend a couple of days visiting her, and we were going to see the Panda Reserve in Chengdu. However, she had an issue with her teaching assignment, and decided to end her contract early, so she came home. I had a choice to make, which was to continue the plan and visit a place where my niece would not be or change the itinerary. Since I was in the Himalayan region, I thought about visiting Nepal. Generally, most people who visit Mt. Everest, travel through Nepal as the entry requirements are significantly easier than Tibet. So, having a couple of extra days, I decided to visit **Kathmandu,** the Capitol of Nepal. As I had spent a week immersed in the Buddhist culture, I would now experience a culture that more closely followed **Hinduism.**

Nepal, while maintaining a significant Buddhist presence, tends to align itself more with Indian culture. My limited knowledge of both cultures, had me thinking it's all the same thing, but it is vastly different. Along the way, I discovered some interesting economic observations as well, that made my time there, that much more enjoyable.

The flight from Lhasa to Kathmandu was just a little over an hour in time, but the scenery was incredible. The flight goes directly over the entire mountain region, and the snowcapped Himalayas was breathtaking. Kathmandu, while still considered mountainous, is considerably lower in altitude than Lhasa, so breathing is much easier, and of course, it was considerably warmer there, and humid. This was a good thing because the next day I was going white water rafting for the first time in my life. After landing, I grabbed a taxi to the hotel, and noticed that traffic was a bit crazy. I would find out very soon why that was, but I just wanted to rest. I stayed at a Fairfield Suites in the **Thamel District** which is where most of the tourism takes place. It was a brand-new hotel opened less than 6 months, and very Westernized. As I mentioned my body was physically exhausted from 2 weeks of hiking Great Walls, and Great Mountains, so I decided to rest when I checked in. That rest turned out to be a 15-hour nap that carried me through till the next morning. Sometimes I just have to stop and take heed of what my 53-year-old body was telling me.

Well rested, the next morning I got an early start for

what would be an exciting white-water raft trip down the **Trishulli River.** I was met by a representative of the rafting company who walked me to the bus that would take me to the raft put in location. It was about 65 miles away, but because of traffic, and road conditions, the trip took 3 ½ hours to get there. When I mentioned that traffic was a bit crazy, it is because Kathmandu has **no traffic lights, or stop signs.** There are no traffic directions of any kind, it is simply a free for all, in a city of over 1 million people. I arrived at the put in location thoroughly frustrated from the travel time, but it was now time for fun. I hooked up my Go Pro, went through the safety course, and we were ready to go. Our raft had about 14 people in it, and we all had a paddle. It was a fun 3-hour ride down the river that included lunch. A great day on the river was followed by a not so great ride back to the hotel. The return took over 4 hours, and I was mentally exhausted returning to the hotel after 9pm.

Surely my next day would be much less stressful, as I booked a mountain bike tour right from the city. No buses, or long rides, just a walk to the location to get the bike, and a way we go. Wrong, the tour started right in the city, and part of the ride to get to the mountain meant riding through downtown Kathmandu. Sometimes people have asked me, what is the craziest thing you have every done, and up to this point, I would always answer "sky diving". But on this day, the new champion was riding a mountain bike through the streets of Kathmandu. Crossing a street was like the old video game **"Frogger"** if you remember that, when trying to go from lane to lane without getting run over.

Once we reached the mountains, the ride was amazing. Some parts of road were very rough, and others not so much. The views of the city and surrounding mountains were breathtaking, as was the humidity when it came to catching your breath. Riding through tiny villages allowed me to see what much of Nepal is all about. Nepal by nature is a very poor country, and in many cases, it is sad to see how little people really have. But I also experienced how happy these people are to have what they do, and I learned something about certain standards that are culturally driven and not economic driven. Now in a country like Nepal, you would anticipate things like bathrooms to be generally terrible, especially in public places. Not one restroom I visited in Nepal was missing soap and running water. Even in the poorest of places, cleanliness is very important, whereas in Tibet, a country considerably wealthier, cleanliness was not important. This was truly an example of what economics drives, and what culture drives.

My final day in Kathmandu was left pretty open, since my flight left late in the evening, I wanted to take it slow and just enjoy the city. The one thing I had to do was go see **Swayambhunath**, aka **The Monkey Temple.** This is a Buddhist temple in the city, that is home to Rhesus Monkeys that run all over the grounds. The monkeys are wild, and as I would find out, very aggressive. I was told that you can feed them bananas, so I bought a couple from a local vendor right there. As I started walking up the steps, you could see them running around, and some would come close and then run away. I thought this would be a good time to shoot a video, so I got the phone ready, and pulled out a banana to peel it. As I was talking in the video describing where I was, I was attacked from behind as a monkey stole the banana from my hand and knocked my phone to the ground. With the video still recording, I picked up the phone, and begin to film the monkey eating the banana. I told him what he did wasn't nice, and at that moment, he turned, looked at me, and flashed a big smile. It was if he knew what I said to him, and the moment that was captured on video was priceless.

As I prepared to leave Nepal, I knew I had one more dragon to slay. I had to make a connection in **Hong Kong** on the way home, so I arranged the connecting flight to provide me a layover of 19 hours. This would provide me just enough time to arrive and take a half day sightseeing tour of Hong Kong Island. Since I was flying Business Class on the return, I was able to leave my carry-on bags at the Cathay Pacific Business Class Lounge, where they would safely store them. The flight from Kathmandu was an overnight flight of 5 hours, but the difference between the two cities is closer to 500 years in progress. Hong Kong is extremely wealthy and developed, with an amazing infrastructure.

I was able to take a train right from the airport to a local hotel where I would meet the tour guide for the afternoon tour. The tour would immediately take us to the tramway to the top of Victoria Peak. Here is where the stunning views of Hong Kong Harbor can be best seen. Going all the way to the top of the observation tower provides a 360-degree view of the harbor and surrounding foothills. This is one of the wealthiest places in all of Hong Kong, and some of the world's wealthiest people live in this area. From the opulent to the working class we traveled the Stanley Marketplace. This is an area of Hong Kong where small shopkeepers have their businesses along the waterway. It was a place for great souvenir shopping, but quite frankly, it was somewhat of a letdown considering many people say it's a must see in Hong Kong. It had been a long day, and it was time for me to head back to the airport for the long Trans-Pacific flight home.

Again, the Cathay Pacific Business Class lounge was amazing, and provided an awesome environment to wait out a 3-hour departure. Having taken an overnight flight and spent the better part of the day in Hong Kong, I was feeling pretty gross. Not to worry as the lounge provided full shower facilities, so I took advantage of that perk. I was able to shower and change into a clean set of clothes, then relax before my flight left in a couple of hours. Flying home in Business Class makes all the difference on a flight of this length. Airline Elite Class services now include private seating pods, where your seat lays flat into a bed, a complete entertainment center, and every amenity you could think of. I had a choice to take Business Class at the beginning of the trip or the end of the trip, and I am very glad I chose the end. I managed to take a 7-hour nap, which made for a very manageable 6-hour flight while awake. It was the perfect end to what can only be described as a perfect trip, and possibly the most liberating trip I have ever taken.

I learned so much about the very mysterious cultures, an appreciation for who Asians are as a people, and also how fortunate we still have things in our country. It was my first experience entering a country that technically is not a free society, but the reality is that much of the bluster surrounding Socialism, is just that, bluster. Make no mistake, the Chinese do not have political freedom, and if constantly redressing your Government is essential to your way of life, you would not like China. However, most of the Chinese people I observed are happy with the stability, and their ability to sustain their families through their hard work.

A final comment on this trip is that I never felt at any time unsafe or nervous, even walking the streets of Tibet at night. The punishments for crime in China are severe, which is why they have a very low crime rate. You can judge the merits of that system against your own morals, and our system of government, but in this day and age where safety is paramount, how refreshing to be free from that burden.

What's Left?

Asia is by far the largest continent in the world, and you might think with so much left to explore, there would be a ton of places that I still want to go. The reality is, while there are some interesting places I would not mind seeing, I have done my bucket list items, so anything in the future would be considered gravy at this point. If I had to choose some specific places or things I would like to see, then I would say:

- Japan, as I would enjoy climbing Mt. Fuji, and seeing some of the historical World War II monuments.

- Singapore and some places In Thailand would also be an interesting trip as a combination of sights and beaches.
- The Temple of Angkor Wat in Cambodia would be a fascinating place to visit.
- I would gladly welcome a return trip to China, Lhasa, and Mt. Everest.

Of course, as I have found out over all these years, a work-related trip can take you to very unsuspecting places, and I will always welcome that.

6–Adventures

It really isn't hard to make a case that all the trips I have highlighted in this book could be considered adventures in my life. But it is just as easy to have adventures in your life without ever traveling anywhere. What happens many times is that we have things that happen in our lives, but we don't view them as actual adventures. If you really look back on things that you have done in the past, you probably could find an adventure or two hidden amongst the everyday routines. It might be something as simple as hiking or biking in a local area or boating on the **Colorado River**. I have indeed done those things with my family, and a whole lot of other outdoor related things. But there have been a few things that standout as being adventures in my life, that just like many of the trips I have taken, have also shaped my perspective about everyday living.

Fear of Heights
ONGOING

For as long as I can remember, I have had a significant fear of heights, or clinically known as **Acrophobia**. Supposedly this fear can be traced to some form of traumatic event in your life, often as a child that brings out the fear when faced in similar situations. I truly do not remember having gone through any event that created this fear, but it is real, and still affects me. A lot of people have asked me how I can fly in airplanes with a fear of heights. The real fear is when you can gauge how far you are from the physical ground. Flying in an airplane does not allow you the ability to see just how far you are from the ground. 35,000 feet in the air, can seem just like 3 feet in the air, so it has never been a problem for me. What has been a problem for me are hotel rooms on high floors with a balcony or overlook viewpoints. Those are two examples of how this fear controls my approach to such situations.

A while back I decided to start facing this fear, but in a controlled way. I didn't want the fear to keep me from doing things that looked fun, but I still had to respect its existence. Up to this point I would never try if they involved heights. So I started small, and managed to work my way up to more challenging things.

It's Only Water

My first adventures with this challenge came when my landing base was water, as I jumped off the cliff into the Caribbean in Jamaica, and as I gave parasailing a go while visiting Club Med. Both of those events I write about in earlier chapters. Telling myself there was nothing to be afraid of because I would land in water certainly helped my mind set, but in the case of parasailing, I remember holding on to the harness with my fists clenched so tight, that when the ride ended, my fingers were stiff, and felt like I was almost afraid to let go. It was clearly a step forward in trying something new and gave me the confidence to try other forms of high adventures. The key was for me to keep challenging myself, because even though I successfully did these things, the fear still remained.

The Ground is Harder than Water

Not all adventures involving height afford you the luxury of a water landing, and therefore I had to take my demons to the next level. There has always been something majestic about a big balloon soaring through the sky that led me to think how I would like to take a hot air balloon flight. There is a Wine Country area near my home in the city of **Temecula**, that in addition to being the home of many wineries, also offers hot air balloon flights in the early morning hours in conjunction with the sunrise. As my 20th Wedding Anniversary was coming up, I thought this would be a nice present for my then wife. It would be something we could each experience for the first time, and also gave me a new way to challenge my fear. The process was quite amazing watching them get the balloon ready for lift off, and soon we were up in the air. My attempt to bring a bit of brevity to the moment by playing the song "**Up, Up and a Way**" by the 5th Dimension, certainly relaxed my nervousness, but didn't impress the other flyers. I had mastered the balloon, and the most memorable thing about that trip was how hot the basket gets because of the gas torch.

Over the last 15 years or so, the activity of ziplining has become popular to the point where many places are now creating zipline parks. The concept of a Tarzan atmosphere swinging from tower to tower was never burning at me, but since I was on this quest, and it was a new activity, I placed it on the list. This adventure would take place on **Catalina Island** in California. The thought of holding on to a bar that glides along a cable from tower to tower sounds pretty scary, and the least safe of anything I had done to date, but I found that the safety precautions were numerous. Before I knew it, I was gliding between towers some 30 to 50 feet above ground looking as if my feet could touch some tree tops. Though it probably would have been in my best interest to continue to look straight ahead, I'll admit I would take a peak down every so often, and even convinced myself this was no big thing. Flying above the tree tops was a big step, but certainly there would be something more adventurous out there that was calling me.

The Sky's The limit

Having literally flown over a million miles in my lifetime, traveling by airplane is not really a big deal for me, and in many cases, it's just what I do. But what about actually flying a plane? My eyesight being what it is has always been a condition that would never allow me to fly a plane professionally, and I knew that early on in my life. So that left my options to fly a small private plane as my only experience, or so I thought.

Back in 2014, I was attending a business meeting with Delta Airlines, who invited us their operations center in Atlanta, GA. In addition to seeing many of the things that make up the airline, we also had the opportunity to ride in a 777-flight simulator. This is what they train their pilots on to certify them for flying certain aircraft. The simulator allows you to program airports to take off and land at, as well as program different kinds of weather, and the process feels so life like it's scary. So now it was my turn to fly this baby, and as sat in the pilot chair, programed the airports and weather I was ready for my moment to shine. Sadly, for my passengers, I skidded off a snowy runway in Salt Lake City for my first landing. Even though my landing was less than stellar, I was in no way daunted for what seemed to be the next logical step, and that was to fly the real thing.

Taking a flying lesson or obtaining a private pilot's license is not a new thing, and people have been doing these things for years. But it was something I had never done and seemed to be the next step in a process that started with me jumping off a stone platform years ago. I receive a Christmas gift from someone that was a Groupon certificate for a 1 hour flying lesson from the local **Riverside airport**. It was not something I had thought about doing, but when the gift came upon me, I thought it was a great idea. So, I booked my flying date, and drove to the airport on a Saturday afternoon. In my mind I had a vision of basically sitting in the cockpit of a small plane and experiencing what flying from this perspective was all about. That was the wrong vision as I would find out that the flying lesson was truly that, flying.

After a brief Ground School session, and a check of the plane, it was take off time. We rolled down the runway and were soon airborne above the City of Riverside. The view for me was spectacular, and of course I had on my professional pilot headphones and could hear cockpit to tower communications. At about 7000 feet, the pilot turned to me, and said "it's all yours". I was a bit surprised that I would actually fly a plane, but I pulled out the yoke, and took his direction, and I was flying this small bird. Increasing the altitude, and making a left turn over Lake Elsinore I remember thinking have I really come this far, to the point of flying a plane through the sky. As we approached landing the pilot took over the controls, and after getting my certificate I said, "I did it". There really wasn't anything else I could do to conquer this fear, so it was finished. Not Quite!

If you had asked me a couple of years ago what was one thing that I would never do, the answer unequivocally would be sky diving. Jumping out of an airplane had absolutely no appeal to me, even as I have been fascinated with flying my entire life. If I have learned one thing from the many things I have tried and done in my life, it is to never say never, and this was just another reminder of that.

The idea of sky diving came into my head at a business meeting when a member of the customer's team showed us a video of her tandem jump for her 50th birthday. It looked amazing, and at that moment a bit of sexism popped into my head and said, "if a 50-year-old lady can do it, why can't this 51-year-old man do it too"? Thus, the idea for me to do a jump was born, along with the notion that I have flown many different ways, I have actually piloted a plane, so the only thing left to do was to jump out of the damn thing. All I had to do was pick a place and a date for which to do the jump, so I chose a place near the Pacific Coast in **Oceanside, CA** on an August morning.

The area where I live in Southern California is almost always sunny during most of the year, however the coast line can often become cloudy. So it was on this day in August, as there was some cloud cover, that would actually delay my 9:00am jump almost 4 hours. I was feeling nervous enough, without dragging the wait out even longer than was necessary, but wait I did. When they gave us the go ahead, I got my instructions and the harness setup, and we were ready to go. My nerves were really heightened at this point as I boarded the plane along with all the other jumpers, and we took off. It took about 20 minutes to reach the 13,000 feet altitude for the jump, but there was still some cloud cover. The first 2 jumpers went out, and at that point, we were told we would have to land because there was too much cloud cover to jump. All that anticipation, and we had to go back to the ground.

After a 30-minute wait on the ground, we were given the all clear, and could try again. The butterflies were even more intense now, but I was determined to do this. We reached the jumping altitude again, and this time I was second in line to go. The first group jumped, and I started to slide down the bench to the open door. My feet were dangling in the open sky 13,000 feet above ground, and I knew there was no backing out. As per instruction, I tilted my head back, and before I knew it, I was freefalling in the sky. The instructor tapped me on the shoulder which was my que to extend my arms out with a thumbs up motion. I was able to see the videographer that I hired to film my jump, and amazingly was able to look directly at her as she was filming.

The freefall lasted 60 seconds, but it seemed much longer, and then the parachute was pulled, and that is probably the most uncomfortable part of the process for men. Up until the parachute, the process for me was not scary at all, as I felt like I was in a wind tunnel. Once the parachute deployed, and I was able to look down and still see we were several thousand feet off the ground, I got nervous, and started holding the straps really tight. Another side effect for me was steering the parachute to the landing spot caused considerable spinning, and by the time I landed, my stomach had become quite queasy. That feeling took away a bit of the initial excitement that I had just jumped out of a plane for the first time and conquered what I considered to be the ultimate fear of height.

I took me a little time to digest what I had just done, but when I saw my video it became a very real thing, and now the big question was, would I do it again? The answer was yes, and 6 months later, at that very same location, on a beautiful clear day, I did my 2nd tandem jump. Not only was the view much better this time, but the fear, and the queasiness of the first jump, was nowhere to be found. I am sure that I will jump again at some point, but there are other first-time adventures I would like to do before I repeat this one. Challenging my fears in this manner was by far the most aggressive I had done to date, and most likely ever will.

The Answer Is
(MAY 1993)

Back in the early 1980's, the board game Trivial Pursuit was sweeping the world off its feet. The game which tested a person's knowledge in a variety of categories, became one of my personal favorites. But like many things in life, everyone seems to have an Achilles heel, and when it came to this game, the category of **Science & Nature** was certainly mine. I just could not answer the science type questions with any regularity, and it was very frustrating to me. Around the same time, the television game show Jeopardy was very popular, and was essentially the same type of game, a test of a person's general knowledge. I have always been someone who seems to store a plethora of useless trivia in my head that can be recalled at a moment's notice. So much so that it was recommended to me that I should go on the game show Jeopardy.

One thing about living in Southern California, there is no shortage of television and movie entertainment being filmed, and the classified sections of the newspapers are filled with requests for game show contestants. So, one day, I decided to give a call to Jeopardy to see about being a contestant, and what I needed to do to get on the show. The process to try out is simple, the task of getting selected is a bit more difficult. A one-hour drive (no traffic, thank you) from home had me at **Sony Pictures Studios** in Culver City, CA where Jeopardy, and other programs are filmed. I was really going to try out for a television game show, and potentially win some money, now that would be an adventure.

After I checked in, a group of 20 people including myself were taken to a sound stage room where we would take the first level test. Depending upon the results of that test would then determine if I moved on in the process. This would be a written test but done very differently. The exact process was outlined as a TV monitor was wheeled in. We would actually simulate a game only Alex Trebek would ask us the questions from a recorded tape, and we had to write down the answers. There would be a total of 50 questions, from a variety of topics, and there was 10 seconds to write down your answer. I felt strong with my geography, history, pop culture, sports and arts categories, but that damn Science stuff reared its ugly head once again. There were several questions asking about mixing various compounds, and periodic elements that I simply had no clue what the answer would be.

After the test was over, they collected everyone's answers, and we were asked to wait while they were graded. If you were fortunate to pass the test, you would then be set up to run a mock game, as if you were on the real show. This is done to see how good your potential stage presence is, and how you would react to actually being filmed for a real show. It wasn't long before all the tests were corrected, and out of 20 people, only 2 people passed the first phase of the test. I was not one of the two, so I would not be on the show, and my day was over.

I was a little disappointed in not making the cut, but it really was an experience to go and try out for one of these shows that you see every day on television. I gave it a go, and it certainly has made for an interesting story to share, and it only strengthened my love of trivia-based games. What I really learned was that the only way I failed in my attempt was if I didn't try at all. Surprisingly, I have never thought about going back and trying again. For me it was a one and done scenario.

I am Jack Ryan
(SEP 1995 – DEC 1995)

I don't think there are many people who would consider a job interview to be an adventure, but there is one for me that had all the makings of it. If you think about the things that make up an adventure, there are a lot of nontraditional things that could be classified as such. During our lifetime we all go through many job interviews, but have you ever had a fantasy job interview? That happened to me in 1995 when I started an interview process with the **CIA.**

It all started innocent enough with a simple advertisement in the Sunday paper. I saw an ad in the "help wanted" section of the newspaper, that the CIA was seeking analysts for their headquarters in **Langley, Virginia.** I have always had an interest in politics, and that is why my degrees are in Political Science with an emphasis on International Relations and Foreign Policy. It was also about this time that the Jack Ryan movies were popular starring Harrison Ford. I wanted to be **Jack Ryan**, and have that kind of job, so I decided to apply for the job.

A couple of months had passed and to be honest, I forgot I even applied for it. Then one day in the mail, I received a letter from **Langley, Virginia** inviting me to complete the application for the analyst position at the CIA. Personally, I thought I had already applied, but as it turns out, I applied for the right to be invited to apply for the job. And boy was this ever an application for the ages that had me question did I really want to do this. Turns out it was a 30-page application that I had to fill out completely. Besides all of the standard information that you would normally find on an application, there was also personal history, residential information for my entire life, and a section of references that I had to provide. Every job I ever worked, every school I had attended, my immediate family members, and personal friends all had to be listed on this application. I also had to write 2 different essays on foreign topics of my choice as a method of evaluating how well my written skills were.

The essays were actually the easy part because of my academic background and interests, and I really felt quite comfortable writing them. But gathering up all that other data was so time consuming and on more than one occasion left me questioning how bad I wanted to be Jack Ryan. Nonetheless, I pushed through the whole process, and finally I had what I considered to be a completed application, all ready to be shipped off to the Central Intelligence Agency. Surely, they could sense my enthusiasm for this job, after all, they are the CIA.

Probably another two plus months went by without hearing anything, and while I did sometimes think about the job, it did slip my mind from time to time. Then one day in December as I was getting ready for the Holidays, I received another letter. This letter was asking me to attend an in-person interview in the Los Angeles area in 2 weeks' time. I actually had been granted a face to face interview with the agency. This led to my re-watching **Patriot Games** for what had become numerous times at this point. What seemed a little strange was that the Interview was going to be held at a local hotel near Los Angeles airport. I didn't give it much thought afterward, but I thought they must have some kind of affiliate office in the area.

On the day of the interview, I had taken the day off from work as I thought I could seriously enhance my chances by being better prepared. I guess I was practicing the old adage **"by failing to prepare, you are preparing to fail"**. I arrived at the interview location, and then realized this was not a personal interview, but that many people had been invited to this level. I felt a bit silly thinking that they wanted to just talk to me, especially after I saw all those people. There were several interviewers there, and they had to get through 100 people. When it was my turn they asked me a couple of questions, mostly about my background and what working for the CIA would mean to me. They also told me that they had received over 5000 applications, and that I was 1 of only 100 people invited to the in-person event.

The next step in the process was to be invited to Langley for additional interviews and in-depth vetting process. Most of the people I had talked to that were there interviewing already had master's degrees, and I was one of the few that was invited that did not already have a master's degree, as I was in the process of completing mine. After having those conversations, and hearing what the interviewers relayed to me, I felt much better about the process, and that I was one of a select few to actually make it this far. But there was still another step to go before I was Jack Ryan.

The wait for the next step was not nearly as long as about 3 weeks from the in-person interview, I received a letter thanking me for my Interest in the CIA, but they would be continuing on with other candidates. You all know the rest of the speech, blah, blah, blah, and the realization that a process that started 4 months earlier had come to an end. I felt bad about not moving on in the process, but I also felt good about how far I went into it. It definitely was an adventure for me, and I honestly felt I put together a strong application, but in the end, I had to accept that I am not Jack Ryan.

What's Next?

While you can never be sure of these unique adventures that may come in and out of your life, I definitely can tell you that there are some physical type adventures that I still want to try. One is the **Jet Pack**, which is offered in the San Diego area close to me. This is the water propelled pack that allows you to fly around and do stunts while attached to a boat. The thought of a bungee jump has crossed my mind, but it is not something that I have been burning to do. Of course, I have said that about a great many other things as well, only to later in life change my mind. I certainly want to spend some more time hiking in some of the local National Parks and continuing to try things that I have never done before. I'd like to think that my motto of **"know your limits but grow your limits"** will continue to guide me the rest of the way.

7- What's Left?

I don't think that anyone will ever reach a point where they say, "I have done everything I wanted to do", or at least I hope I don't reach that point. I have tried to live my life with the notion that **"my reach should always exceed my grasp".** With that thought in mind, there are still 2 continents that I have not been to, and 2 Natural wonders of the world that I want to see, as well as numerous other places that are more local.

Africa

I have gone back and forth about this continent and what I would really like to see as my first visit. The overall prevailing theme is that I want to take some kind of a safari, but not necessarily a long drawn out one. I think it would be incredible to witness large game animals in their natural habitat. A friend of mine has taken several safaris in different African countries, and the pictures she has shared are amazing. Lions lying down on the road as if they were neighborhood cats, and giraffes walking past the hotel terrace during breakfast. I would also like to see the **Pyramids in Egypt,** and much of the other biblical history, but sadly the political situations in that area have not calmed to level that I feel comfortable traveling to. **Climbing Mt. Sinai** is something I would love to do, and I will try, but again, much will depend on the political climate.

There are also some countries in Sub Saharan Africa that I had thought about visiting such as **South Africa**. Once an extremely wealthy and developed country, South Africa has evolved into a country filled with political corruption and is often identified as a place that is not as safe as it once was. One place that I would absolutely love to see are the Seychelles Islands. East of Kenya in the Indian Ocean, the Seychelles have what seem to be the most perfect of beaches. Whichever place I get to first, visiting Africa is a must on the to do list.

Antarctica
(THE BOTTOM OF THE WORLD)

Visiting the continent of Antarctica is probably one of the most difficult places, and perhaps the least desired by people to see. There are many reasons that this continent remains the most elusive of all 7. Because of its location, 98% of Antarctica is covered in ice, and is, home to the South Pole. The continent is claimed by no country and is home to many countries research outposts. Because of this neutral status, there are no official entry requirements to go to Antarctica, no passports required etc. Also, there is no scheduled airline service, and the only flights in and out are government aircraft or specialized charter flights associated with tourism. Many countries have their own research stations down there, and many private companies also participate in the research of this frozen continent, but very little is still understood about it. America has outposts down there, and the McMurdow station is its own self-contained city. Hard to believe when my only reference to Antarctica up till now has been John Carpenter's movie **"The Thing"**.

While it offers what some say are the most pristine places on this planet, only a true handful of people have probably ever seen them. Because it is generally only visited by the truly adventurous, the costs of a trip there are extremely high, starting in the neighborhood of about $10,000. If you are really up for an adventure and you would like to visit the actual South Pole, you will need close to $50,000 per person for a guided trip there. The most logical route for me to get to Antarctica would be through the countries of Argentina or Chile. The majority of people who visit the Antarctica Peninsula, do so via cruise vessels. The larger vessels provide more Western style comfort, but are restricted to how close they can sail into the islands. There are some smaller cruise vessels that visit the tip of Antarctica that allow for shore excursions, but those excursions are a bit more rustic in nature than most people would like. My feeling is that if I reach 6 continents, and the only one left is Antarctica, I will find a way to visit. Even if I just touch my foot on the ice, and see one penguin, I will make it happen.

The Two Natural Wonders of The World.

Yes, there are really seven, however with my trip to Alaska seeing the Aurora in 2019, that increased my viewing of these awesome wonders to five. This really started out as an unexpected quest in that my travel to places just happened to be where some of them are located. I knew the Grand Canyon was one, and we did the family trip there. Traveling to Australia, most people make it a point to visit the Great Barrier Reef, usually for diving or tropical weather. When I went to Rio de Janeiro, I did not think about the harbor as being one of the seven. It wasn't until I decided to go to Everest and started studying the seven that I realized I had another mission that could run parallel with my continents quest. Seeing the Aurora Borealis on my trip to Alaska has only increased my desire for the final two.

1. Victoria Falls (Zambia/Zimbabwe, Africa)

I know very little about the history of the Victoria Falls, other than it is larger than Niagara Falls, and it divides a border of Zambia and Zimbabwe. It is one of the seven, and if my travels take me to Sub-Saharan Africa for perhaps a safari, I will coordinate an opportunity to see this place. The pictures of it look amazing, and there are many hiking tours that include this area, so it definitely is on the radar.

2. Paricutin Volcano (Mexico

Much like visiting Antarctica, if I have the other six in my back pocket, I will do my best to appear here, but it is unlikely I will visit this location before I visit the other places on the list. Given that I have very little travel to Mexico from a business perspective, this would definitely be more of a leisure pursuit.

3. The Tomb of The Unknown Adventure

This is the place that I do not know about yet, because I have not learned of it, or heard about it from word of mouth, or been there related to a business trip. As I wrote throughout this book, many places I got to see were unexpected joys that came about through circumstances that I would not have envisioned myself. Much like the Washington DC memorial "**Tomb of The Unknown Soldier,** this one is precious to me, because it is the adventure for me that is known only to God.

The Three Natural Wonders of The World

Yes, there are really seven, however with my trip to Mt. Everest in 2017, that increased my visits to these awesome places to four. This really started out as an unexpected quest in that my travel to places just happened to be where some of them are located. I knew the Grand Canyon was one, and we did the family trip there. Traveling to Australia, most people make it a point to visit the Great Barrier Reef, usually for diving or tropical weather. When I went to Rio de Janeiro, I did not think about the harbor as being one of the seven. It wasn't until I decided to go to Everest and started studying the seven that I realized I had another mission that could run parallel with my continents quest.

Aurora Borealis (Northern Lights)

Seems that everyone knows what this phenomenon is, but not many people actually get to see it. My strategy to visit here is most likely to go up through **Alaska**, North of Fairbanks perhaps, and view it from there. Although it can be viewed from many different countries that are North of the Arctic Circle, Alaska is the most logical path for me. The most optimal time to see the lights are during the Winter months when daylight is scarce, and so are positive temperatures. This will be an opportunity to tie in a scenic wonder, and an adventure activity as I plan to do my best **Yukon Cornelius** and go for a dog sled ride. I have always wanted to yell **"mush"** as I am pulled on a sled.

Victoria Falls (Zambia/Zimbabwe, Africa)

I know very little about the history of the Victoria Falls, other than it is larger than Niagara Falls, and it divides a border of **Zambia and Zimbabwe.** It is one of the seven, and if my travels take me to Sub-Saharan Africa for perhaps a safari, I will coordinate an opportunity to see this place. The pictures of it look amazing, and there are many hiking tours that include this area, so it definitely is on the radar.

Paricutin Volcano (Mexico)

Much like visiting Antarctica, if I have the other six in my back pocket, I will do my best to appear here, but it is unlikely I will visit this location before I visit the other places on the list. Given that I have very little travel to Mexico from a business perspective, this would definitely be more of a leisure pursuit.

The Tomb of The Unknown Adventure

This is the place that I do not know about yet, because I have not learned of it, or heard about it from word of mouth, or been there related to a business trip. As I wrote throughout this book, many places I got to see were unexpected joys that came about through circumstances that I would not have envisioned myself. Much like the Washington DC memorial **"Tomb of The Unknown Soldier,** this one is precious to me, because it is the adventure for me that is known only to God.

Landing

"Ladies and gentlemen, we will be arriving at our destination in just a few minutes. At this time, it would be a good idea to power down all electronic equipment, and safely stow it away for landing".

So, as we come to the end of this particular adventure, I hope I have created a picture in your mind about the wonders of traveling to different places? Seeing different sites, meeting fascinating people, and discovering new things about the world and myself, has given me a perspective that would have passed me by had I not taken advantage of the opportunities presented to me. By creating a life filled with memories as opposed to stuff, I have something that can never be taken from me, never becomes worn out, and never grows old. There is nothing unique that allowed me to do most of these things, just a desire to seek more with the time I have been given.

As I pointed out in a previous chapter, one does not have to travel around the globe to seek or have adventures. There are many adventures right in front of you if you truly wish to find them. It is my hope that the stories I shared with you, will be the incentive to find your adventure, plan to do it, and then follow through. With that, as

I have throughout the book, I will leave you with a quote from the movie **"Scarface"**, **"The World is Yours"**, which happened to be a tagline used at that time by Pan American Airways.

"The aircraft has come to a complete stop, you are now free to move about the world"